CHEMICAL ENGINEERING

VOLUME 5

Solutions to the Problems in Chemical Engineering Volume 2

CHEMICAL ENGINEERING

VOLUME 5

J. M. COULSON AND J. F. RICHARDSON

Solutions to the Problems in Chemical Engineering Volume 2

By

J. R. BACKHURST AND J. H. HARKER

University of Newcastle upon Tyne

PERGAMON PRESS

OXFORD · NEW YORK · TORONTO · SYDNEY · PARIS · FRANKFURT

√ 6174932I

CHEMISTRY

U.K.	Pergamon Press Ltd., Headington Hill Hall, Oxford OX3 0BW, England
U.S.A.	Pergamon Press Inc., Maxwell House, Fairview Park, Elmsford, New York 10523, U.S.A.
CANADA	Pergamon of Canada, Suite 104, 150 Consumers Road, Willowdale, Ontario M2J 1P9, Canada
AUSTRALIA	Pergamon Press (Aust.) Pty. Ltd., PO Box 544, Potts Point, N.S.W. 2011, Australia
FRANCE	Pergamon Press SARL, 24 rue des Ecoles, 75240 Paris, Cedex 05, France
FEDERAL REPUBLIC OF GERMANY	Pergamon Press GmbH, 6242 Kronberg-Taunus, Pferdstrasse 1, Federal Republic of Germany

First edition 1979

British Library Cataloguing in Publication Data

Chemical engineering.
Vol. 5: Solutions to the problems in 'Chemical engineering', volume 2
1. Chemical engineering
I. Coulson, John Metcalfe   II. Richardson, John Francis
III. Backhurst, John Rayner   IV. Harker, John Hadlett
660.2   TP155   78-40923

ISBN 0-08-022951-4 (Hardcover)
ISBN 0-08-022952-2 (Flexicover)

Printed in Great Britain by Mather Bros (Printers) Limited, Preston

CONTENTS

PREFACE

IN THE preface to the first edition of *Chemical Engineering*, Volume 4, we quoted the following paragraph written by Coulson and Richardson in their preface to the first edition of *Chemical Engineering*, Volume 1:

'We have introduced into each chapter a number of worked examples which we believe are essential to a proper understanding of the methods of treatment given in the text. It is very desirable for a student to understand a worked example before tackling fresh practical problems himself. Chemical engineering problems require a numerical answer, and it is essential to become familiar with the different techniques so that the answer is obtained by systematic methods rather than by intuition.'

It is with these aims in mind that we have prepared Volume 5, which gives our solutions to the problems in the third edition of *Chemical Engineering*, Volume 2. The material is grouped in sections corresponding to the chapters in that volume and the present book is complementary in that extensive reference has been made to the equations and sources of data in Volume 2 at all stages. The book has been written concurrently with the revision of Volume 2 and SI units have been used.

In many ways these problems are more taxing and certainly longer than those in Volume 4, which gives the solutions to problems in Volume 1, and yet they have considerable merit in that they are concerned with real fluids and, more importantly, with industrial equipment and conditions. For this reason we hope that our efforts will be of interest to the professional engineer in industry as well as to the student, who must surely take some delight in the number of tutorial and examination questions which are attempted here.

We are again delighted to acknowledge the help we have received from Professors Coulson and Richardson in so many ways. The former has the enviable gift of providing the minimum of data on which to frame a simple key question, which illustrates the crux of the problem perfectly, whilst the latter has in a very gentle and yet thorough way corrected our mercifully few mistakes and checked the entire work. Our colleagues at the University of Newcastle upon Tyne have again helped us, in many cases unwittingly, and for this we are grateful.

Newcastle upon Tyne, 1978   J. R. BACKHURST
                                                                         J. H. HARKER

SECTION 1

PARTICULATE SOLIDS

Problem 1.1

The size analysis of a powdered material on a weight basis is represented by a straight line from 0% weight at 1 μm particle size to 100% weight at 101 μm particle size. Calculate the mean surface diameter of the particles constituting the system.

Solution

From equation 1.15 the surface mean diameter is given by:

$$d_s = 1/\Sigma(x_1/d_1)$$

Since the size analysis is represented by the continuous line:

$$d = 100x + 1 \qquad\qquad \text{(Fig. 1.7)}$$

$$d_s = 1 \Big/ \int_0^1 dx/d$$

$$= 1 \Big/ \int_0^1 dx/(100x + 1)$$

$$= 100/\ln 101$$

$$= 21 \cdot 7 \, \mu\text{m}$$

Problem 1.2

The equations giving the number distribution curve for a powdered material are $dn/dd = d$ for the size range 0 to 10 μm and $dn/dd = 100{,}000/d^4$ for the size range 10 to 100 μm. Sketch the number, surface, and weight distribution curves. Calculate the surface mean diameter for the powder.

Explain briefly how the data for the construction of these curves would be obtained experimentally.

1

Solution

For the range, $d = 0 - 10\,\mu m$,

$$\frac{dn}{dd} = d$$

On integration, $n = 0{\cdot}5d^2 + c_1$ (i)

where c_1 is the constant of integration.
    For the range, $d = 10 - 100\,\mu m$,

$$\frac{dn}{dd} = 10^5 d^{-4}$$

On integration, $n = c_2 - 0{\cdot}33 \times 10^5 d^{-3}$ (ii)

when $d = 0$, $n = 0$, and from (i) $c_1 = 0$

when $d = 10\,\mu m$, in (i) $n = 0{\cdot}5 \times 10^2 = 50$

    in (ii) $50 = c_2 - 0{\cdot}33 \times 10^5 \times 10^{-3}$

and $c_2 = 83{\cdot}0$

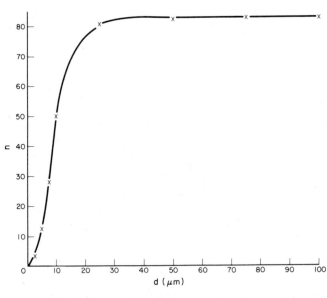

FIG. 1a

Thus for $d = 0 - 10\,\mu m$, $n = 0{\cdot}5d^2$

and for $d = 10 - 100\,\mu m$, $n = 83{\cdot}0 - 0{\cdot}33 \times 10^5 d^{-3}$

Using these equations, values of n are obtained as follows:

$d\,(\mu m)$	n	$d\,(\mu m)$	n
0	0	10	50·0
2·5	3·1	25	80·9
5·0	12·5	50	82·7
7·5	28·1	75	82·9
10·0	50·0	100	83·0

and these data are plotted in Fig. 1a.

From this plot, values of d are obtained for chosen values of n and hence n_1 and d_1 obtained for each increment of n. Values of $n_1 d_1^2$ and $n_1 d_1^3$ are calculated and the totals obtained. The surface area of the particles in the increment is then given by:

$$s_1 = n_1 d_1^2 / \Sigma n_1 d_1^2$$

and s is then found as Σs_1. Similarly the weight of the particles $x = \Sigma x_1$ where:

$$x_1 = n_1 d_1^3 / \Sigma n_1 d_1^3$$

The results are given as follows:

n	d	n_1	d_1	$n_1 d_1^2$	$n_1 d_1^3$	s_1	s	x_1	x
0	0								
		20	3·1	192	596	0·014	0·014	0·001	0·001
20	6·2								
		20	7·6	1155	8780	0·085	0·099	0·021	0·022
40	9·0								
		10	9·5	903	8573	0·066	0·165	0·020	0·042
50	10·0								
		10	10·7	1145	12250	0·084	0·249	0·029	0·071
60	11·4								
		5	11·75	690	8111	0·057	0·300	0·019	0·090
65	12·1								
		5	12·85	826	10609	0·061	0·361	0·025	0·115
70	13·6								
		2	14·15	400	5666	0·029	0·390	0·013	0·128
72	14·7								
		2	15·35	471	7234	0·035	0·425	0·017	0·145
74	16·0								
		2	16·75	561	9399	0·041	0·466	0·022	0·167
76	17·5								
		2	18·6	692	12870	0·057	0·517	0·030	0·197
78	19·7								
		2	21·2	890	18877	0·065	0·582	0·044	0·241
80	22·7								
		1	24·1	581	14000	0·043	0·625	0·033	0·274
81	25·5								
		1	28·5	812	23150	0·060	0·685	0·055	0·329
82	31·5								
		1	65·75	4323	284240	0·316	1·000	0·670	1·000
83	100								
				13641	424355				

Values of s and x are plotted as functions of d in Fig. 1b.
Surface mean diameter,

$$d_s = \Sigma(n_1 d_1^3)/\Sigma(n_1 d_1^2) = 1/\Sigma(x_1/d_1)$$

$$= \int d^3 \, dn \bigg/ \int d^2 \, dn$$

For $0 < d < 10 \, \mu m$, $dn = d \, dd$

For $10 < d < 100 \, \mu m$, $dn = 10^5 d^{-4} \, dd$

$$\therefore \quad d_s = \left(\int_0^{10} d^4 \, dd + \int_{10}^{100} 10^5 d^{-1} \, dd \right) \bigg/ \left(\int_0^{10} d^3 \, dd + \int_{10}^{100} 10^5 d^{-2} \, dd \right)$$

$$= ([d^5/5]_0^{10} + 10^5 [\ln d]_{10}^{100}) / ([d^4/4]_0^{10} + 10^5 [-d^{-1}]_{10}^{100})$$

$$= (2 \times 10^4 + 2 \cdot 303 \times 10^5)/(2 \cdot 5 \times 10^3 + 9 \times 10^3)$$

$$= \underline{\underline{21 \cdot 8 \, \mu m}}$$

The size range of a material is determined by sieving for relatively large particles and by sedimentation methods for particles which are too fine for sieving. The use of such a method is described in Chapter 1, Volume 2.

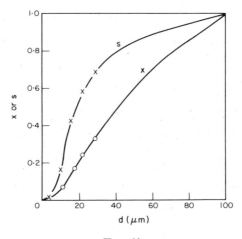

FIG. 1b

Problem 1.3

The fineness characteristic of a powder on a cumulative basis is represented by a straight line from the origin to 100% undersize at particle size 50 μm. If the powder is initially dispersed uniformly in a column of liquid, calculate the proportion by weight which remains in suspension at a time interval from commencement of settling such that a 40 μm particle would fall the total height of the column.

Solution

For flow under streamline conditions, the velocity is proportional to the diameter squared. Hence the time taken for a 40 μm particle to fall a height h m is:

$$h/40^2k \text{ s}$$

During this time, a particle of diameter d μm has fallen a distance:

$$kd^2h/40^2k = hd^2/40^2 \text{ m}$$

The proportion of particles of size d which are still in suspension $= (1 - d^2/40^2)$ and the fraction by mass of particles which are in suspension

$$= \int_0^{40} (1 - d^2/40^2) \, dw$$

Since $dw/dd = 1/50$, the mass fraction:

$$= (1/50) \int_0^{40} (1 - d^2/40^2) \, dd$$

$$= (1/50)[d - d^3/4800]_0^{40} = 0.533$$

or $\underline{53.3\%}$ of the particles remain in suspension

Problem 1.4

In a mixture of quartz, specific gravity 2·65, and galena, specific gravity 7·5, the sizes of the particles range from 0·0052 to 0·025 mm.

On separation in a hydraulic classifier under free settling conditions, three fractions are obtained, one consisting of quartz only, one a mixture of quartz and galena, and one of galena only. What are the ranges of sizes of particles of the two substances in the mixed portion?

Solution

Use is made of equation 3.17, which may be written as:

$$u = kd^2(\rho_s - \rho)$$

For large galena, $u = k \times 0.025^2 (7.5 - 1.0) = 0.004\,06k$ mm/s

For small galena, $u = k \times 0.0052^2 (7.5 - 1.0) = 0.000\,175k$ mm/s

For large quartz, $u = k \times 0.025^2 (2.65 - 1.0) = 0.001\,03k$ mm/s

For small quartz, $u = k \times 0.0052^2 (2.65 - 1.0) = 0.000\,045k$ mm/s

If the time of settling was such that particles with a velocity equal to 0·001 03k mm/s settled, then the bottom product would contain quartz. This is not so and hence the maximum size of galena particles still in suspension is given by:

$$0.001\,03k = kd^2\,(7.5 - 1.0)$$

$\therefore$
$$d = 0.0126\,\text{mm}$$

Similarly if the time of settling was such that particles with a velocity equal to $0.000\,175k$ mm/s did not start to settle, then the top product would contain galena. This is not the case and hence the minimum size of quartz in suspension is given by:

$$0.000\,175k = kd^2\,(2.65 - 1.0)$$

or
$$d = 0.0103\,\text{mm}$$

It may therefore be concluded that, assuming streamline conditions, the fraction of material in suspension, that is containing quartz *and* galena, is made up of particles of sizes in the range:

$$\underline{\underline{0.0103-0.0126\,\text{mm}}}$$

Problem 1.5

It is desired to separate into two pure fractions a mixture of quartz and galena of a size range from 0.015 mm to 0.065 mm by the use of a hindered settling process. What is the minimum apparent density of the fluid that will give this separation? How will the viscosity of the bed affect the minimum required density?
(Density of galena = 7500 kg/m³, density of quartz = 2650 kg/m³.)

Solution

Assuming that the shapes of the galena and quartz particles are similar, then from equation 1.31, the required density of fluid for viscous conditions is given by:

$$(0.065/0.015) = [(7500 - \rho)/(2650 - \rho)]^{0.5}$$

or
$$\rho = 2377\,\text{kg/m}^3$$

From equation 1.33, the required density for fully turbulent conditions is given by:

$$(0.065/0.015) = (7500 - \rho)/(2650 - \rho)$$

and
$$\rho = 1196\,\text{kg/m}^3$$

Thus the minimum density of the fluid to effect the separation is $\underline{1196\,\text{kg/m}^3}$. This assumes that fully turbulent conditions prevail. As the viscosity is increased, the value of the Reynolds group will decrease and the required density will rise to the value of $\underline{2377\,\text{kg/m}^3}$ necessary for viscous conditions.

Problem 1.6

Write a short essay explaining the circumstances in which a particle size distribution would be determined by microscopical measurement or by sedimentation in a liquid. State the characteristics of these two methods of measurement.

The following table gives the size distribution of a dust as measured by the microscope. Convert these figures to obtain the distribution on a weight basis, and calculate the specific surface, assuming spherical particles of specific gravity 2·65.

Size range in μm	Number of particles in range
0–2	2000
2–4	600
4–8	140
8–12	40
12–16	15
16–20	5
20–24	2

Solution

The determination of a particle size distribution by microscopic and sedimentation techniques is discussed in section 1.2.2 to which reference should be made. In simple terms, sedimentation would be adopted for very small particles which are too fine for classification by either sieving or visual examination.

From equation 1.4, the mass fraction of particles of size d_1 is given by:

$$x_1 = n_1 k_1 d_1^3 \rho_s$$

where k_1 is a constant, n_1 the number of particles of size d_1, and ρ_s the density of the particles $= 2650 \text{ kg/m}^3$.

$\Sigma x_1 = 1$ and hence the weight fraction is

$$x_1 = n_1 k_1 d_1^3 \rho_s / \Sigma nkd^3 \rho_s$$

In this case:

d	n	$kd^3 n \rho_s$	x
1	200	5,300,000k	0·011
3	600	42,930,000k	0·090
6	140	80,136,000k	0·168
10	40	106,000,000k	0·222
14	15	109,074,000k	0·229
18	5	77,274,000k	0·162
22	2	56,434,400k	0·118
		$\Sigma = 477,148,400k$	$\Sigma = 1.0$

The surface mean diameter is given by equation 1.14:

$$d_s = \Sigma (n_1 d_1^3)/\Sigma (n_1 d_1^2)$$

The working is as follows:

d	n	nd^2	nd^3
1	2000	2000	2,000
3	600	5400	16,200
6	140	5040	30,240
10	40	4000	40,000
14	15	2940	41,160
18	5	1620	29,160
22	2	968	21,296
		$\Sigma = 21{,}968$	$\Sigma = 180{,}056$

$$\therefore \qquad d_s = 180{,}056/21{,}968 = 8\cdot20\,\mu m$$

This is the size of particle with the same specific surface as the mixture.

Volume of a particle $8\cdot20\,\mu m$ diameter $= (\pi/6) \times 8\cdot20^3 = 288\cdot7\,\mu m^3$.

Surface area of a particle of $8\cdot20\,\mu m$ diameter $= \pi \times 8\cdot20^2 = 211\cdot2\,\mu m^2$ and hence specific surface $= 211\cdot2/288\cdot7 = 0\cdot731\,\mu m^2/\mu m^3$

or
$$0\cdot731 \times 10^6 \, m^2/m^3$$

Problem 1.7

The performance of a solids mixer has been assessed by calculating the variance occurring in the weight fraction of a component amongst a selection of samples withdrawn from the mixture. The quality was tested at intervals of 30 s and the results are:

Sample variance	0·025	0·006	0·015	0·018	0·019
Mixing time (s)	30	60	90	120	150

If the component analysed is estimated to represent 20% of the mixture by weight and each of the samples removed contained approximately 100 particles, comment on the quality of the mixture produced and present the data in graphical form showing the variation of mixing index with time.

Solution

For a completely unmixed system, $s_0^2 = p(1-p)$ (equation 1.23)

$$= 0\cdot20(1-0\cdot20) = 0\cdot16$$

For a completely random mixture, $s_r^2 = p(1-p)/n$ (equation 1.22)

$$= 0\cdot20(1-0\cdot20)/100 = 0\cdot0016$$

The degree of mixing M is given by equation 1.24 as:

$$M = (s_0^2 - s^2)/(s_0^2 - s_r^2)$$

In this case,
$$M = (0.16 - s^2)/(0.16 - 0.0016)$$
$$= 1.01 - 6.313s^2$$

The calculated data are therefore:

t (s)	30	60	90	120	150
s^2	0.025	0.006	0.015	0.018	0.019
M	0.852	0.972	0.915	0.896	0.890

These data are plotted in Fig. 1c and it is clear that the degree of mixing is a maximum at $t = 60$ s.

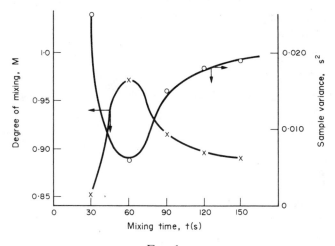

FIG. 1c

SECTION 2

SIZE REDUCTION OF SOLIDS

Problem 2.1

A material is crushed in a Blake jaw crusher and the average size of particle reduced from 50 mm to 10 mm with the consumption of energy at the rate of 13·0 kW/(kg/s). What will be the consumption of energy needed to crush the same material of average size 75 mm to an average size of 25 mm:

(a) Assuming Rittinger's law applies?
(b) Assuming Kick's law applies?

Which of these results would be regarded as being more reliable and why?

Solution

(a) *Rittinger's Law*

This is given by equation 2.2:

$$E = K_R f_c \left(\frac{1}{L_2} - \frac{1}{L_1} \right)$$

or

$$13·0 = K_R f_c (\tfrac{1}{10} - \tfrac{1}{50})$$

∴

$$K_R f_c = 13·0 \times 50/4 = 162·5 \text{ kW s/kg mm}$$

Thus the energy required to crush 75 mm material to 25 mm is given by:

$$E = 162·5(\tfrac{1}{25} - \tfrac{1}{75}) = \underline{\underline{4·33 \text{ kW/(kg/s)}}}$$

(b) *Kick's Law*

This is given by equation 2.3:

$$E = K_K f_c \ln (L_1/L_2)$$

or

$$13·0 = K_K f_c \ln (50/10)$$

∴

$$K_K f_c = 13·0/1·609 = 8·08 \text{ kW/(kg/s)}$$

Thus the energy required to crush 75 mm material to 25 mm is given by:

$$E = 8·08 \ln (75/25) = \underline{\underline{8·88 \text{ kW/(kg/s)}}}$$

The size range involved may be classified as coarse crushing, and because Kick's law more closely relates to the energy required to effect elastic deformation before fracture occurs, this would be taken as giving the more accurate result.

Problem 2.2

A crusher was used to crush a material whose compressive strength was 22·5 MN/m². The size of the feed was *minus* 50 mm, *plus* 40 mm, and the power required was 13·0 kW/(kg/s). The screen analysis of the product was as follows:

Size of aperture (mm)	Per cent of product
Through 6·00	100
On 4·00	26
On 2·00	18
On 0·75	23
On 0·50	8
On 0·25	17
On 0·125	3
Through 0·125	5

What would be the power required to crush 1 kg/s of a material of compressive strength 45 MN/m² from a feed *minus* 45 mm, *plus* 40 mm to a product of average size 0·50 mm?

Solution

The first part of the working is to obtain a dimension representing the mean size of the product. Using Bond's method of taking the size of opening through which 80% of the material will pass, a value of just over 4·00 mm is indicated by the data. Alternatively, calculations may be made as follows:

Size of aperture (mm)	Mean d_1 (mm)	n_1	nd_1	nd_1^2	nd_1^3	nd_1^4
6·00						
	5·00	0·26	1·3	6·5	32·5	162.5
4·00						
	3·00	0·18	0·54	1·62	4·86	14·58
2·00						
	1·375	0·23	0·316	0·435	0·598	0·822
0·75						
	0·67	0·08	0·0536	0·0359	0·0241	0·0161
0·50						
	0·37	0·17	0·0629	0·0233	0·0086	0·003 19
0·25						
	0·1875	0·03	0·0056	0·001 05	0·000 20	0·000 037
0·125						
	0·125	0·05	0·006 25	0·000 78	0·000 098	0·000 012
			2·284	8·616	37·991	177·92

From equation 1.11, the weight mean diameter,

$$d_v = \Sigma n_1 d_1^4 / \Sigma n_1 d_1^3$$

$$= 177 \cdot 92 / 37 \cdot 991 = 4 \cdot 683 \text{ mm}$$

From equation 1.14, the surface mean diameter,

$$d_s = \Sigma n_1 d_1^3 / \Sigma n_1 d_1^2$$

$$= 37 \cdot 991 / 8 \cdot 616 = 4 \cdot 409 \text{ mm}$$

From equation 1.18, the linear mean diameter,

$$d_1 = \Sigma n_1 d_1^2 / \Sigma n_1 d_1$$

$$= 8 \cdot 616 / 2 \cdot 284 = 3 \cdot 772 \text{ mm}$$

From equation 1.19, the mean linear diameter,

$$d_1' = \Sigma n_1 d_1 / \Sigma n_1$$

$$= 2 \cdot 284 / 1 \cdot 0 = 2 \cdot 284 \text{ mm}$$

In the present situation, which is concerned with power consumption per unit mass, the weight mean diameter is probably of the greatest relevance. For the purposes of calculation, a mean value of 4·0 mm will be used, which agrees with the value obtained by Bond's method. For coarse crushing, Kick's law may be used as follows:

Case 1: mean diameter of feed = 45 mm
   mean diameter of product = 4 mm
   energy consumption = 13·0 kW/(kg/s)
   compressive strength = 22·5 MN/m^2

In equation 2.3,

$$13 \cdot 0 = K_K \times 22 \cdot 5 \ln (45/4)$$

and $$K_K = 13 \cdot 0 / 54 \cdot 4 = 0 \cdot 239 \text{ kW/(kg/s)(MN/m}^2)$$

Case 2: mean diameter of feed = 42·5 mm
   mean diameter of product = 0·50 mm
   compressive strength = 45 MN/m^2

∴ $$E = 0 \cdot 239 \times 45 \ln (42 \cdot 5 / 0 \cdot 50) = 0 \cdot 239 \times 199 \cdot 9$$

$$= 47 \cdot 8 \text{ kW/(kg/s)}$$

or, for a feed of 1 kg/s,   the energy required = 47·8 kW

Problem 2.3

A crusher in reducing limestone of crushing strength 70 MN/m^2 from 6 mm diameter average size to 0·1 mm diameter average size requires 9 kW. The same machine is used to crush dolomite at the same rate of output from 6 mm diameter average size to a product which consists of 20% with an average diameter of 0·25 mm, 60% with an average diameter of 0·125 mm, the balance having an average diameter of 0·085 mm. Estimate the power required to drive the crusher, assuming that the crushing strength of the dolomite is 100 MN/m² and that crushing follows Rittinger's law.

Solution

The weight mean diameter of the crushed dolomite is calculated as:

n_1	d_1	$n_1 d_1^3$	$n_1 d_1^4$
0·20	0·250	0·003 125	0·000 78
0·60	0·125	0·001 172	0·000 146
0·20	0·085	0·000 123	0·000 011
		0·004 42	0·000 937

and from equation 1.11:

$$d_v = \Sigma n_1 d_1^4 / \Sigma n_1 d_1^3 = 0.000\,937/0.004\,42$$

$$= 0.212\,\text{mm}$$

For Case 1: $E = 9·0\,\text{kW}$
$f_c = 70·0\,\text{MN/m}^2$
$L_1 = 6·0\,\text{mm}$
$L_2 = 0·1\,\text{mm}$

and in equation 2.2:

$$9·0 = K_R \times 70·0\left(\frac{1}{0·1} - \frac{1}{6·0}\right)$$

or

$$K_R = 0·013\,\text{kW mm/(MN/m}^2)$$

For Case 2: $f_c = 100·0\,\text{MN/m}^2$
$L_1 = 6·0\,\text{mm}$
$L_2 = 0·212\,\text{mm}$

Hence

$$E = 0·013 \times 100·0\left(\frac{1}{0·212} - \frac{1}{6·0}\right)$$

$$= \underline{\underline{5·9\,\text{kW}}}$$

Problem 2.4

If crushing rolls 1 m in diameter are set so that the crushing surfaces are 12·5 m apart and the angle of nip is 31°, what is the maximum size of particle which should be fed to the rolls?

If the actual capacity of the machine is 12% of the theoretical, calculate the throughput in kg/s when running at 2·0 Hz if the working face of the rolls is 0·4 m long and the feed weighs 2500 kg/m^3.

Solution

The particle size may be obtained from equation 2.6:

$$\cos \alpha = (r_1 + b)/(r_1 + r_2)$$

In this case, $2\alpha = 31°$ and $\cos \alpha = 0·964,$

$$b = 12·5/2 = 6·25 \text{ mm}   \text{ or }   0·00625 \text{ m}$$

$$r_1 = 1·0/2 = 0·5 \text{ m}$$

∴ $0·964 = (0·5 + 0·00625)/(0·5 + r_2)$

∴ $r_2 = 0·025 \text{ m}   \text{ or }   \underline{25 \text{ mm}}$

Cross-sectional area for flow $= (0·0125 \times 0·4) = 0·005 \text{ m}^2$ and the volumetric flow rate $= (2·0 \times 0·005) = 0·010 \text{ m}^3/\text{s}$.

The actual throughput $= (0·010 \times 12/100) = 0·0012 \text{ m}^3/\text{s}$ or

$$(0·0012 \times 2500) = \underline{3·0 \text{ kg/s}}$$

Problem 2.5

A crushing mill reduces limestone from a mean particle size of 45 mm to a product:

Size (mm)	Per cent
12·5	0·5
7·5	7·5
5·0	45·0
2·5	19·0
1·5	16·0
0·75	8·0
0·40	3·0
0·20	1·0

and in so doing requires 21 kJ/kg of material crushed.

Calculate the power required to crush the same material at the same rate, from a feed having a mean size of 25 mm to a product with a mean size of 1 mm.

Solution

The mean size of the product may be obtained as:

n_1	d_1	$n_1 d_1^3$	$n_1 d_1^4$
0·5	12·5	3906	48,828
7·5	7·5	3164	23,731
45·0	5·0	5625	28,125
19·0	2·5	296·9	742·2
16·0	1·5	54·0	81·0
8·0	0·75	3·375	2·531
3·0	0·40	0·192	0·0768
1·0	0·20	0·008	0·0016
		13,049	101,510

and from equation 1.11, the weight mean diameter,

$$d_v = \Sigma n_1 d_1^4 / \Sigma n_1 d_1^3 = 101{,}510/13{,}049 = 7\cdot78 \text{ mm}$$

Kick's law will be used as the present case may be regarded as coarse crushing.

Case 1: $E = 21$ kJ/kg

$\quad L_1 = 45$ mm

$\quad L_2 = 7\cdot8$ mm

In equation 2.3:

$$21 = K_K f_c \ln(45/7\cdot8)$$

and $\qquad\qquad K_K f_c = 11\cdot98 \text{ kJ/kg}$

Case 2: $L_1 = 25$ mm

$\quad L_2 = 1\cdot0$ mm

$\therefore \qquad\qquad E = 11\cdot98 \ln(25/1\cdot0)$

$$= \underline{\underline{38\cdot6 \text{ kJ/kg}}}$$

Problem 2.6

A ball mill 1·2 m in diameter is being run at 0·80 Hz; it is found that the mill is not working satisfactorily. Would you suggest any modification in the conditions of operation?

Solution

The critical angular velocity is given by equation 2.8:

$$w_c = \sqrt{(g/r)}$$

In this equation, r is the radius of the mill less that of the particle. For small particles, $r = 0.6$ m and hence:

$$w_c = \sqrt{(9.81/0.6)} = 4.04 \text{ rad/s}$$

$$\text{The actual speed} = (2\pi \times 0.80) = 5.02 \text{ rad/s}$$

and hence it may be concluded that the speed of rotation is too high and the balls are being carried round in contact with the sides of the mill with little relative movement or grinding taking place.

The optimum speed of rotation is $0.5w_c$ to $0.75w_c$, say $0.6w_c$ or

$$(0.6 \times 4.04) = 2.42 \text{ rad/s}$$

This is equivalent to $(2.42/2\pi) = 0.39$ Hz or, in simple terms, the speed of rotation should be halved.

Problem 2.7

3 kW has to be supplied to a machine crushing material at the rate of 0.3 kg/s from 12.5 mm cubes to a product having the following sizes:

$$\begin{array}{ll} 80\% & 3.175 \text{ mm} \\ 10\% & 2.5 \text{ mm} \\ 10\% & 2.25 \text{ mm} \end{array}$$

What would be the power which would have to be supplied to this machine to crush 0.3 kg/s of the same material from 7.5 mm cube to 2.0 mm cube?

Solution

The weight mean diameter may be calculated as:

n_1	d_1	$n_1 d_1^3$	$n_1 d_1^4$
0.8	3.175	25.605	81.295
0.1	2.5	1.563	3.906
0.1	2.25	1.139	2.563
		28.307	87.763

and from equation 1.11:

$$d_v = \Sigma n_1 d_1^4 / \Sigma n_1 d_1^3 = 87.763/28.307 = 3.100 \text{ mm}$$

(Using Bond's approach, the mean diameter is clearly 3.175 mm.)

For the size ranges involved, the crushing may be considered as intermediate and Bond's law will be used.

Case 1: $E = 3/0{\cdot}3 = 10\,kW/(kg/s)$

$\qquad L_1 = 12{\cdot}5\,mm$

$\qquad L_2 = 3{\cdot}1\,mm$

$\therefore$ In equation 2.4: $\qquad q = L_1/L_2 = 4{\cdot}03$

and $\qquad\qquad\qquad E = 2C\sqrt{(1/L_2)}(1 - 1/q^{0{\cdot}5})$

$\qquad\qquad\qquad 10 = 2C\sqrt{(1/3{\cdot}1)}(1 - 1/4{\cdot}03^{0{\cdot}5})$

$\qquad\qquad\qquad\quad = 2C \times 0{\cdot}568 \times 0{\cdot}502$

$\therefore \qquad\qquad\qquad C = 17{\cdot}54\,kW\,mm^{0{\cdot}5}/(kg/s)$

Case 2: $L_1 = 7{\cdot}5\,mm$

$\qquad L_2 = 2{\cdot}0\,mm$

$\qquad q = (7{\cdot}5/2{\cdot}0) = 3{\cdot}75$

$\therefore \qquad\qquad\qquad E = 2 \times 17{\cdot}54(1/2{\cdot}0)(1 - 1/3{\cdot}75^{0{\cdot}5})$

$\qquad\qquad\qquad\quad = (35{\cdot}08 \times 0{\cdot}707 \times 0{\cdot}484)$

$\qquad\qquad\qquad\quad = 12{\cdot}0\,kW/(kg/s)$

For a feed of $0{\cdot}3\,kg/s$, the power required $= (12{\cdot}0 \times 0{\cdot}3)$

$$= \underline{\underline{3{\cdot}60\,kW}}$$

MOTION OF PARTICLES IN A FLUID

Problem 3.1

A finely ground mixture of galena and limestone in the proportion of 1 to 4 by weight is subjected to elutriation by an upwards current of water flowing at 5 mm/s. Assuming that the size distribution for each material is the same, and as shown by the following table, estimate the percentage of galena in the material carried away and in the material left behind. Take the absolute viscosity of water as 1 mN s/m² and use Stokes' equation.

Diameter (μm)	20	30	40	50	60	70	80	100
Percentage weight of undersize	15	28	48	54	64	72	78	88

Specific gravity of galena = 7·5; specific gravity of limestone = 2·7.

Solution

It is necessary to determine the size of particle which has a settling velocity equal to that of the upward flow of fluid, 5 mm/s.

Taking the largest particle, $d = 100 \times 10^{-6} = 10^{-4}$ m

$$\therefore \qquad Re' = 5 \times 10^{-3} \times 10^{-4} \times 1000/l \times 10^{-3} = 0.5$$

Thus for the bulk of particles the flow will be within region a and the settling velocity is given by Stokes' equation (3.17):

$$u = (d^2 g / 18 \mu)(\rho_s - \rho)$$

For a particle of galena settling at 5 mm/s,

$$5 \times 10^{-3} = (d^2 \times 9.81/18 \times 10^{-3})(7500 - 1000)$$

$$= 3.54 \times 10^6 d^2$$

and $\qquad d = 3.76 \times 10^{-5}$ m $= 37.6 \,\mu$m

For a particle of limestone settling at 5 mm/s,

$$5 \times 10^{-3} = (d^2 \times 9.81/18 \times 10^{-3})(2700 - 1000)$$

$$= 9.27 \times 10^5 d^2$$

and $\qquad d = 7.35 \times 10^{-5}$ m $= 73.5 \,\mu$m

Thus particles of galena less than 37·6 μm and particles of limestone less than 73·5 μm will be removed in the water stream.

Interpolation of the data given shows that 43% of the galena and 74% of the limestone will be removed in this way.

In 100 kg feed, there is 20 kg galena and 80 kg limestone.

Therefore galena removed = $(20 \times 0\cdot43) = 8\cdot6$ kg, leaving 11·4 kg, and limestone removed = $(80 \times 0\cdot74) = 59\cdot2$ kg, leaving 20·8 kg.

Hence in *material removed*,

$$\text{percentage galena} = 8\cdot6 \times 100/(8\cdot6 + 59\cdot2) = \underline{\underline{12\cdot7\%}}$$

and in *material remaining*,

$$\text{percentage galena} = 11\cdot4 \times 100/(11\cdot4 + 20\cdot8) = \underline{\underline{35\cdot4\%}}$$

Problem 3.2

Calculate the terminal velocity of a steel ball, 2 mm diameter (density 7870 kg/m^3) in oil (density 900 kg/m^3, viscosity 50 mN s/m^2).

Solution

For a sphere $(R'_0/\rho u_0^2)\,Re'^2_0 = (2d^3/3\mu^2)\rho(\rho_s - \rho)g$ (equation 3.28)

$$= (2 \times 0\cdot002^3/3 \times 0\cdot05^2)\,900\,(7870 - 900)\,9\cdot81$$

$$= 131\cdot3$$

$$\log_{10} 131\cdot3 = 2\cdot118$$

From Table 3.2, $\log_{10} Re'_0 = 0\cdot833$

∴ $Re'_0 = 6\cdot80$

∴ $u_0 = 6\cdot80 \times 0\cdot05/(900 \times 0\cdot002)$

$$= \underline{\underline{0\cdot189\ \text{m/s}}}$$

Problem 3.3

What will be the settling velocity of a spherical particle 0·40 mm diameter in an oil of specific gravity 0·82 and viscosity 10 mN s/m^2? The specific gravity of steel is 7·87.

Solution

For a sphere

$$(R'_0/\rho u_0^2)\,Re'^2_0 = (2d^3\rho/3\mu^2)(\rho_s - \rho)g \qquad \text{(equation 3.28)}$$

Chemical Engineering

$$= [2 \times 0.0004^3 \times 820/3(10 \times 10^{-3})^2](7870 - 820)\,9.81$$

$$= 24.2$$

$$\log_{10} 24.2 = 1.384$$

$$\therefore \qquad \log_{10} Re'_0 = 0.222 \quad \text{(from Table 3.2)}$$

$$\therefore \qquad Re'_0 = 1.667$$

$$\therefore \qquad u_0 = 1.667 \times 10 \times 10^{-3}/(0.0004 \times 820)$$

$$= \underline{0.051 \text{ m/s}} \quad (51 \text{ mm/s})$$

Problem 3.4

What will be the settling velocities of mica plates 1 mm thick and ranging in area from 6 to 600 mm^2 in an oil of specific gravity 0·82 and viscosity 10 mN s/m^2? The specific gravity of mica is 3·0.

Solution

	Smallest particles	Largest particles
A'	$= 6 \times 10^{-6}\,\text{m}^2$	$6 \times 10^{-4}\,\text{m}^2$
d_p	$= \sqrt{(4 \times 6 \times 10^{-6}/\pi)} = 2.76 \times 10^{-3}\,\text{m}$	$\sqrt{(4 \times 6 \times 10^{-4}/\pi)} = 2.76 \times 10^{-2}\,\text{m}$
d_p^3	$2.103 \times 10^{-8}\,\text{m}^3$	$2.103 \times 10^{-5}\,\text{m}^3$
volume	$6 \times 10^{-9}\,\text{m}^3$	$6 \times 10^{-7}\,\text{m}^3$
k'	0.285	0.0285

$$(R'_0/\rho u^2)\,Re'^2_0 = (4k'/\mu^2\pi)(\rho_s - \rho)\,\rho d_p^3 g \qquad\qquad \text{(equation 3.36)}$$

$$= (4 \times 0.285/\pi \times 0.01^2)(3000 - 820)\,820 \times 2.103 \times 10^{-8} \times 9.81$$

$$= 1340 \text{ for smallest particle and } 134{,}000 \text{ for largest particle}$$

	Smallest particles	Largest particles
$\log_{10}(R'_0/\rho u^2)\,Re'^2_0$	3.127	5.127
$\log_{10} Re'_0$	1.581	2.857 (from Table 3.2)
Correction from Table 3.4	-0.038	-0.300 (estimated)
Corrected $\log_{10} Re'_0$	1.543	2.557
Re'_0	34.9	361
u	$\underline{0.154 \text{ m/s}}$	$\underline{0.159 \text{ m/s}}$

Thus it is seen that all the mica particles settle at approximately the same velocity.

Problem 3.5

A material of specific gravity 2·5 is fed to a size separation plant where the separating fluid is water which rises with a velocity of 1·2 m/s. The upward vertical component of the velocity of the particles is 6·0 m/s. How far will an approximately spherical particle, 6 mm diameter, rise relative to the walls of the plant before it comes to rest in the fluid?

Solution

Initial velocity of particle relative to fluid, $v = (6\cdot0 - 1\cdot2) = 4\cdot8$ m/s

$$Re' = (6 \times 10^{-3} \times 4\cdot8 \times 1000)/1 \times 10^{-3} = 28,800$$

When the particle has been retarded to such a velocity that $Re' = 500$, the minimum value for which equation 3.76 is applicable,

$$\dot{y} = (4\cdot8 \times 500/28,800) = 0\cdot083 \text{ m/s}$$

When Re' is greater than 500, the relation between the displacement of the particle y and time t is:

$$y = -(1/c)\ln(\cos fct - (v/f)\sin fct) \qquad \text{(equation 3.76)}$$

where

$$c = (0\cdot33/d)(\rho/\rho_s) = (0\cdot33/6 \times 10^{-3})(1000/2500) = 22\cdot0 \qquad \text{(equation 3.62)}$$

$$f = \sqrt{\{[d(\rho_s - \rho)g]/0\cdot33\rho\}} = \sqrt{[(6 \times 10^{-3} \times 1500 \times 9\cdot81)/0\cdot33 \times 1000]} = 0\cdot517$$
$$\text{(equation 3.75)}$$

$$v = -4\cdot8 \text{ m/s}$$

Thus

$$y = -(1/22\cdot0)\ln[\cos 0\cdot517 \times 22t + (4\cdot8/0\cdot517)\sin 0\cdot517 \times 22t]$$

$$= -0\cdot0455\ln(\cos 11\cdot37t + 9\cdot28\sin 11\cdot37t)$$

$$\therefore \dot{y} = -0\cdot0455\left(\frac{-11\cdot37\sin 11\cdot37t + 9\cdot28 \times 11\cdot37\cos 11\cdot37t}{\cos 11\cdot37t + 9\cdot28\sin 11\cdot37t}\right)$$

$$= -\frac{0\cdot517(9\cdot28\cos 11\cdot37t - \sin 11\cdot37t)}{\cos 11\cdot37t + 9\cdot28\sin 11\cdot37t}$$

The time taken for the velocity of the particle relative to the fluid to fall from 4·8 m/s to 0·083 m/s is given by:

$$-0\cdot083 = 0\cdot517(9\cdot28\cos 11\cdot37t - \sin 11\cdot37t)/(\cos 11\cdot37t + 9\cdot28\sin 11\cdot37t)$$

i.e. $\qquad \cos 11\cdot37t + 9\cdot28\sin 11\cdot37t = -6\cdot23\sin 11\cdot37t + 57\cdot8\cos 11\cdot37t$

i.e. $\qquad 56\cdot8\cos 11\cdot37t = 15\cdot51\sin 11\cdot37t$

$\therefore \qquad \sin 11\cdot37t = 3\cdot66\cos 11\cdot37t$

Squaring $1 - \cos^2 11\cdot37t = 13\cdot4\cos^2 11\cdot37t$

$\therefore$ $\cos 11\cdot37t = 0\cdot264$

$\therefore$ $\sin 11\cdot37t = \sqrt{(1-0\cdot264^2)} = 0\cdot965$

The distance moved by the particle relative to the fluid during this period is therefore given by:

$$y = -0\cdot0455 \ln (0\cdot264 + 9\cdot28 \times 0\cdot965)$$

$$= -0\cdot101 \text{ m}$$

If equation 3.76 were applied for a relative velocity down to zero, the time taken for the particle to come to rest would be given by:

$$9\cdot28 \cos 11\cdot37t = \sin 11\cdot37t$$

Squaring, $1 - \cos^2 11\cdot37t = 86\cdot1 \cos^2 11\cdot37t$

$\therefore$ $\cos 11\cdot37t = 0\cdot107$

and $\sin 11\cdot37t = \sqrt{(1-0\cdot107^2)} = 0\cdot994$

The corresponding distance the particle moves relative to the fluid is then given by:

$$y = -0\cdot0455 \ln (0\cdot017 + 9\cdot28 \times 0\cdot994)$$

$$= -0\cdot102 \text{ m}$$

i.e. the particle moves only a very small distance with a velocity of less than 0·083 m/s.

If form drag were neglected for all velocities less than 0·083 m/s, the distance moved by the particle would be given by equation 3.53:

$$y = (b/a)t + (v/a) - (b/a^2) + \left(\frac{b}{a^2} - \frac{v}{a}\right)e^{-at}$$

and $\dot{y} = \dfrac{b}{a} - \left(\dfrac{b}{a} - v\right)e^{-at}$

where $a = 18\mu/d^2\rho_s = (18 \times 0\cdot001)/(0\cdot006^2 \times 2500)$ (equation 3.51)

$$= 0\cdot20$$

$b = (1 - \rho/\rho_s)g = (1 - 1000/2500)9\cdot81 = 5\cdot89$ (equation 3.52)

$\therefore$ $b/a = 29\cdot43$

and $v = -0\cdot083 \text{ m/s}$

Thus $y = 29\cdot43t - \left(\dfrac{0\cdot083}{0\cdot20} + \dfrac{29\cdot43}{0\cdot20}\right)(1 - e^{-0\cdot20t})$

$$= 29\cdot43t - \frac{29\cdot51}{0\cdot20}(1 - e^{-0\cdot20t})$$

$\therefore$ $\dot{y} = 29\cdot43 - 29\cdot51 e^{-0\cdot20t}$

When the particle comes to rest in the fluid, $y = 0$ and

$$e^{-0.20t} = 29.43/29.51$$

$$\therefore \qquad t = 0.0141\,s$$

The corresponding distance moved by the particle is given by:

$$y = 29.43 \times 0.0141 - (29.51/0.20)(1 - e^{-0.20 \times 0.0141})$$

$$= 0.41442 - 0.41550 = -0.00108\,m$$

Thus, whether the resistance force is calculated by equation 3.9 or equation 3.13, the particle moves a negligible distance with a velocity relative to the fluid of less than 0.083 m/s. Further, the time is also negligible and thus the fluid also has moved only a very small distance.

It can therefore be taken that the particle moves through 0.102 m before it comes to rest in the fluid. The time taken for the particle to move this distance is given by equation 3.79, on the assumption that the drag force corresponds to that given by equation 3.13. The time is therefore given by:

$$\cos 11.37t = 0.264 \qquad\qquad \text{(equation 3.79)}$$

$$\therefore \qquad 11.37t = 1.304$$

and

$$t = 0.115\,s$$

The distance travelled by the fluid in this time $= (1.2 \times 0.115) = 0.138$ m. Thus the total distance moved by the particle relative to the walls of the plant

$$= (0.102 + 0.138) = \underline{\underline{0.240\,m}}$$

Problem 3.6

A spherical glass particle is allowed to settle freely in water. If the particle starts initially from rest and if the value of the Reynolds number (Re') with respect to the particle is 0.1 when it has attained its terminal velocity, calculate:

(a) the distance travelled before the particle reaches 90% of its terminal velocity, and
(b) the time which has elapsed when the acceleration of the particle is one-hundredth of its initial value.

Solution

When $Re' < 0.2$, the terminal velocity is given by equation 3.17:

$$u_0 = (d^2 g/18\mu)(\rho_s - \rho)$$

Taking the densities of glass and water as 2750 and 1000 kg/m³ respectively and the viscosity of water as 0.001 Ns/m²,

$$u_0 = (9.81d^2/18 \times 0.001)(2750 - 1000)$$

$$= 9.54 \times 10^5 d^2\,m/s$$

The Reynolds number $Re' = 0.1$ and substituting for u,

$$d(9.54 \times 10^5 d^2) \times 1000/0.001 = 0.1$$

$\therefore$ $\qquad d = 4.76 \times 10^{-5}$ m

Now $\qquad a = 18\mu/d^2\rho_s = (18 \times 0.001)/(4.76 \times 10^{-5})^2 \times 2750 = 2889/s$

and $\qquad b = (1 - \rho/\rho_s)g = (1 - 1000/2750)9.81 = 6.24$ m/s²

In equation 3.53

$$y = \frac{b}{a}t + \frac{v}{a} - \frac{b}{a^2} + \left(\frac{b}{a^2} - \frac{v}{a}\right)e^{-at}$$

In this case $v = 0$ and on differentiating:

$$\dot{y} = \frac{b}{a}(1 - e^{-at})$$

or, since $b/a = u$, the terminal velocity,

$$\dot{y} = u(1 - e^{-at})$$

When $\dot{y} = 0.9u$, $\qquad 0.9 = (1 - e^{-2889t})$

$\therefore$ $\qquad\qquad 2889t = 2.303$

$\therefore$ $\qquad\qquad t = 8.0 \times 10^{-4}$ s

Thus in equation 3.53:

$$y = (6.24 \times 8.0 \times 10^{-4}/2889) - (6.24/2889^2) + (6.24/2889^2)\exp(-2889 \times 8.0 \times 10^{-4})$$

$$= 1.73 \times 10^{-6} - 7.52 \times 10^{-7} + 7.513 \times 10^{-8}$$

$$= 1.053 \times 10^{-6}\text{ m} \quad \text{or} \quad \underline{\underline{1.05\text{ mm}}}$$

From equation 3.50:

$$\ddot{y} = b - a\dot{y}$$

At the start of the fall, $\dot{y} = 0$ and the initial acceleration, $\ddot{y} = b$.
    When $\ddot{y} = 0.01b$,

$$0.01b = b - a\dot{y}$$

or $\qquad\qquad \dot{y} = (0.89 \times 6.24)/2889$

$$= 0.002\,14\text{ m/s}$$

$\therefore$ $\qquad 0.002\,14 = (6.24/2889)(1 - e^{-2889t})$

$$2889t = 4.605$$

$\therefore$ $\qquad\qquad \underline{\underline{t = 0.0016\text{ s}}}$

Problem 3.7

In a hydraulic jig, a mixture of two solids is separated into its components by subjecting an aqueous slurry of the material to a pulsating motion and allowing the particles to settle for a series of short-time intervals such that they do not attain their terminal falling velocities. It is desired to separate materials of specific gravities 1·8 and 2·5 whose particle size ranges from 0·3 to 3 mm diameter. It may be assumed that the particles are approximately spherical and that Stokes' law is applicable. Calculate approximately the maximum time interval for which the particles may be allowed to settle so that no particle of the less dense material falls a greater distance than any particle of the denser material.

Viscosity of water $= 1 \, \text{mN s/m}^2$.

Solution

For Stokes' law to apply, $Re' < 0·2$ and resistance is due to skin friction only. Equation 3.53 may be used:

$$y = \frac{b}{a}t + \frac{v}{a} - \frac{b}{a^2} + \left(\frac{b}{a^2} - \frac{v}{a}\right)e^{-at}$$

or, assuming the initial velocity $v = 0$,

$$y = \frac{b}{a}t - \frac{b}{a^2} + \frac{b}{a^2}e^{-at}$$

where $b = (1 - \rho/\rho_s)g$ and $a = 18\mu/d^2\rho_s$.

For small particles of the dense material,

$$b = (1 - 1000/2500)9·81 = 5·89 \, \text{m/s}^2$$

$$a = (18 \times 0·001)/(0·3 \times 10^{-3})^2 2500 = 80/s$$

For large particles of the light material,

$$b = (1 - 1000/1800)9·81 = 4·36 \, \text{m/s}^2$$

$$a = (18 \times 0·001)/(3 \times 10^{-3})^2 1800 = 1·11/s$$

In order that these particles should fall the same distance, in equation 3.53:

$$(5·89/80)t - (5·89/80^2)(1 - e^{-80t}) = (4·36/1·11)t - (4·36/1·11^2)(1 - e^{-1·11t})$$

$$\therefore \qquad 3·8504t + 3·5316 e^{-1·11t} - 0·000\,92\,e^{-80t} = 3·5307$$

and solving by trial and error,

$$\underline{\underline{t = 0·01 \, \text{s}}}$$

Problem 3.8

Two spheres of equal terminal velocity settle in water starting from rest at the same horizontal level. How far apart vertically will the particles be when they have both reached their terminal falling velocities? Assume Stokes' law is valid and then check the assumption.

Data:

	Density (kg/m^3)	Viscosity (mN s/m^2)	Diameter (μm)
Particle 1	1500	—	40
Particle 2	3000	—	—
Water	1000	1	—

Solution

Assuming Stokes' law is valid, the terminal velocity is given by equation 3.17:

$$u_t = (d^2 g/18\mu)(\rho_s - \rho)$$

For particle 1,

$$u_t = [(40 \times 10^{-6})^2 \times 9\cdot81/(18 \times 1 \times 10^{-3})](1500 - 1000)$$

$$= 4\cdot36 \times 10^{-4}\,\text{m/s}$$

Since particle 2 has an equal terminal velocity:

$$4\cdot36 \times 10^{-4} = [(d_2^2 \times 9\cdot81)/(18 \times 1 \times 10^{-3})](3000 - 1000)$$

From which,

$$d_2 = 2 \times 10^{-5}\,\text{m} \quad \text{or} \quad 20\,\mu\text{m}$$

From equation 3.51:

$$a = 18\mu/d^2\rho_s$$

and for particle 1,

$$a_1 = 18 \times 1 \times 10^{-3}/(40 \times 10^{-6})^2 \times 1500 = 7\cdot5 \times 10^3/\text{s}$$

and for particle 2,

$$a_2 = 18 \times 1 \times 10^{-3}/(20 \times 10^{-6})^2 \times 3000 = 1\cdot5 \times 10^4/\text{s}$$

From equation 3.52:

$$b = (1 - \rho/\rho_s)g$$

and for particle 1, $b_1 = (1 - 1000/1500)9\cdot81 = 3\cdot27\,\text{m/s}^2$

and for particle 2, $b_2 = (1 - 1000/3000)9\cdot81 = 6\cdot54\,\text{m/s}^2$

The initial velocity of both particles, $v = 0$ and from equation 3.53:

$$y = \frac{b}{a}t - \frac{b}{a^2} + \frac{b}{a^2}e^{-at}$$

Differentiating,

$$\dot{y} = \frac{b}{a}(1 - e^{-at})$$

or, from equation 3.17:

$$\dot{y} = u_t(1 - e^{-at})$$

When $\dot{y} = u_t$, the terminal velocity, it is not possible to solve for t and hence $\dot{y}$ will be taken as $0.99u_t$.

For particle 1:

$$0.99 \times 4.36 \times 10^{-4} = (4.36 \times 10^{-4})[1 - \exp(-7.5 \times 10^3 t)]$$

$\therefore$

$$t = 6.14 \times 10^{-4}\,\mathrm{s}$$

The distance travelled in this time is given by equation 3.53:

$$y = (3.27/7.5 \times 10^3)6.14 \times 10^{-4} - [3.27/(7.5 \times 10^3)^2]$$
$$\times [1 - \exp(-7.5 \times 10^3 \times 6.14 \times 10^{-4})]$$
$$= 2.10 \times 10^{-7}\,\mathrm{m}$$

For particle 2:

$$0.99 \times 4.36 \times 10^{-4} = (4.36 \times 10^{-4})[1 - \exp(-1.5 \times 10^4 t)]$$

$\therefore$

$$t = 3.07 \times 10^{-4}\,\mathrm{s}$$

and

$$y = (6.54/1.5 \times 10^4)3.07 \times 10^{-4} - [6.54/(1.5 \times 10^4)^2]$$
$$\times [1 - \exp(-1.5 \times 10^4 \times 3.07 \times 10^{-4})]$$
$$= 1.03 \times 10^{-7}\,\mathrm{m}$$

Particle 2 reaches its terminal velocity after $3.07 \times 10^{-4}\,\mathrm{s}$ and it then travels at $4.36 \times 10^{-4}\,\mathrm{m/s}$ for a further $(6.14 \times 10^{-4} - 3.07 \times 10^{-4}) = 3.07 \times 10^{-4}\,\mathrm{s}$ during which time it travels a further $(3.07 \times 10^{-4} \times 4.36 \times 10^{-4}) = 1.338 \times 10^{-7}\,\mathrm{m}$.

Thus the total distance moved by particle $1 = 2.10 \times 10^{-7}\,\mathrm{m}$

and the total distance moved by particle $2 = (1.03 \times 10^{-7} + 1.338 \times 10^{-7})$

$$= 2.368 \times 10^{-7}\,\mathrm{m}$$

The distance apart when both particles have attained their terminal velocities

$$= (2.368 \times 10^{-7} - 2.10 \times 10^{-7}) = \underline{\underline{2.68 \times 10^{-8}\,\mathrm{m}}}$$

For Stokes' law to be valid, Re' must be less than 0.2 when the terminal velocities are attained:

for particle 1,

$$Re = (40 \times 10^{-6} \times 4.36 \times 10^{-4} \times 1500)/(1 \times 10^{-3}) = 0.026$$

and for particle 2,

$$Re = (20 \times 10^{-6} \times 4.36 \times 10^{-4} \times 3000)/(1 \times 10^{-3}) = 0.026$$

and the law does apply.

Problem 3.9

The size analysis of a powder is carried out by sedimentation in a vessel having the sampling point 180 mm below the liquid surface. If the viscosity of the liquid is $1 \cdot 2\,\text{mN s/m}^2$, and the densities of the powder and liquid are 2650 and $1000\,\text{kg/m}^3$ respectively, determine the time which must elapse before any sample will exclude particles larger than $20\,\mu\text{m}$.

If turbulent conditions occur when the Reynolds number is greater than $0 \cdot 2$, what is the approximate maximum size of particle to which Stokes' law can be applied under the above conditions?

Solution

The problem involves obtaining the time taken for a $20\,\mu\text{m}$ particle to fall below the sampling point, i.e. 180 mm. Assuming that skin friction is the only resistance, equation 3.53 may be used, taking the initial velocity $v = 0$:

$$y = bt/a - b/a^2(1 - e^{-at})$$

where

$$b = g(1 - \rho/\rho_s) = 9 \cdot 81(1 - 1000/2650) = 6 \cdot 108\,\text{m/s}^2$$

$$a = 18\mu/d^2\rho_s = (18 \times 1 \cdot 2 \times 10^{-3})/(20 \times 10^{-6})^2 \times 2650$$

$$= 20{,}377/\text{s}$$

In this case $y = 180\,\text{mm} = 0 \cdot 180\,\text{m}$

$\therefore$

$$0 \cdot 180 = (6 \cdot 108/20{,}377)\,t - (6 \cdot 108/20{,}377^2)(1 - e^{-20{,}377t})$$

$$= 0 \cdot 0003t + 1 \cdot 4071 \times 10^{-8}\,e^{-20{,}377t}$$

Ignoring the exponential term as being negligible,

$$t = 0 \cdot 180/0 \cdot 0003$$

$$= \underline{\underline{600\,\text{s}}}$$

The velocity is given by differentiating equation 3.53:

$$\dot{y} = \frac{b}{a}(1 - e^{-at})$$

When $t = 600\,\text{s}$,

$$\dot{y} = [(6 \cdot 108d^2 \times 2650)/(18 \times 0 \cdot 0012)]\{1 - \exp[-(18 \times 0 \cdot 0012 \times 600)/d^2 \times 2650]\}$$

$$= 7 \cdot 49 \times 10^5 d^2[1 - \exp(-4 \cdot 89 \times 10^{-3}d^{-2})]$$

For $Re' = 0 \cdot 2$,

$$d(7 \cdot 49 \times 10^5 d^2)[1 - \exp(-4 \cdot 89 \times 10^{-3}d^{-2})] \times 2650/0 \cdot 0012 = 0 \cdot 2$$

$\therefore$

$$1 \cdot 65 \times 10^{12}d^3[1 - \exp(-4 \cdot 89 \times 10^{-3}d^{-2})] = 0 \cdot 2$$

As d will be small, the exponential term is negligible and

$$d^3 = 1 \cdot 212 \times 10^{-13}$$
$$d = 5 \cdot 46 \times 10^{-5}\,\text{m} = \underline{\underline{54 \cdot 6\,\mu\text{m}}}$$

Problem 3.10

Calculate the distance a spherical particle of lead shot of diameter (d) 0·1 mm will settle in a glycerol/water mixture before it reaches 99% of its terminal falling velocity.
Density of lead = 11,400 kg/m³.
Density of liquid = 1000 kg/m³.
Viscosity of liquid (μ) = 10 mN s/m².

Assume that the resistance force can be calculated from Stokes' law and is equal to $3\pi\mu du$, where u is the velocity of the particle relative to the liquid.

Solution

The terminal velocity, when Stokes' law applies, is given by:

$$\tfrac{1}{6}\pi d^3(\rho_s-\rho)g = 3\pi\mu du$$

$$\therefore \qquad u_0 = \frac{d^2 g}{18\mu}(\rho_s-\rho) \qquad\qquad \text{(equation 3.17)}$$

$$= \frac{d^2\rho_s}{18\mu}g(1-\rho/\rho_s)$$

$$= b/a$$

where $\qquad b = g(1-\rho/\rho_s) = 9{\cdot}81(1-1000/11{,}400) = 8{\cdot}95 \text{ m/s}^2$

and $\qquad a = 18\mu/d^2\rho_s = (18 \times 10 \times 10^{-3})/(0{\cdot}1 \times 10^{-3})^2 11{,}400$

$$= 1579/\text{s}$$

$$\therefore \qquad u_0 = 8{\cdot}95/1579 = 5{\cdot}67 \times 10^{-3} \text{ m/s}$$

When 99% of this velocity is attained,

$$\dot{y} = 0{\cdot}99 \times 5{\cdot}67 \times 10^{-3}$$

$$= 5{\cdot}61 \times 10^{-3} \text{ m/s}$$

Assuming the initial velocity v is zero, equation 3.53 may be differentiated to give:

$$\dot{y} = (b/a)(1-e^{-at})$$

$$\therefore \qquad 5{\cdot}61 \times 10^{-3} = 5{\cdot}67 \times 10^{-3}(1-e^{-1579t})$$

and $\qquad\qquad t = 0{\cdot}0029 \text{ s}$

Substituting in equation 3.53:

$$y = (b/a)t - (b/a^2)(1-e^{-at})$$

$$= (5{\cdot}67 \times 10^{-3} \times 0{\cdot}0029) - (5{\cdot}67 \times 10^{-3}/1579)(1-e^{-1579 \times 0{\cdot}0029})$$

$$= 1{\cdot}644 \times 10^{-5} - 3{\cdot}59 \times 10^{-6} \times 9{\cdot}89 \times 10^{-1}$$

$$= 1{\cdot}29 \times 10^{-5} \text{ m} \quad \text{or} \quad \underline{\underline{0{\cdot}013 \text{ mm}}}$$

Problem 3.11

Find the weight of a sphere of material of specific gravity 7·5 which falls with a steady velocity of 0·6 m/s in a large deep tank of water.

Solution

In equation 3.33:

$$\frac{R'_0}{\rho u_0^2} Re'_0{}^{-1} = \frac{2\mu g}{3\rho^2 u_0^3} (\rho_s - \rho)$$

Taking the density and viscosity of water as 1000 kg/m^3 and 0.001 N s/m^2 respectively,

$$(R'_0/\rho u_0^2)/Re'_0 = [(2 \times 0.001 \times 9.81)/(3 \times 1000^2 \times 0.6^3)](7500 - 1000)$$

$$= 0.000\,197$$

$$\therefore \qquad\qquad \log_{10}(R'_0/\rho u_0^2)/Re'_0 = \overline{4}\cdot296$$

From Table 3.3, $\log_{10} Re'_0 = 3.068$

$\therefore \qquad\qquad\qquad Re'_0 = 1169.5$

$\therefore \qquad\qquad\qquad d = (1169.5 \times 0.001)/(0.6 \times 1000)$

$$= 0.001\,95 \text{ m} \equiv 1.95 \text{ mm}$$

and the weight $= \pi d^3 \rho_s/6$

$$= \pi \times 0.001\,95^3 \times 7500/6$$

$$= 2.908 \times 10^{-5} \text{ kg} \quad \text{or} \quad \underline{\underline{0.029 \text{ g}}}$$

Problem 3.12

Two ores, of specific gravities 3·7 and 9·8, are to be separated in water by a hydraulic classification method. If the particles are all of approximately the same shape and each is sufficiently large for the drag force to be proportional to the square of the velocity in the fluid, calculate the maximum ratio of sizes which can be separated if the particles attain their terminal velocities. Explain why a wider range of sizes can be separated if the time of settling is so small that the particles do not reach their terminal velocities.

Obtain an explicit expression for the distance through which a particle will settle in a given time if it starts from rest and if the resistance force is proportional to the square of the velocity. The acceleration period is to be taken into account.

Solution

If the total drag force is proportional to the square of the velocity, when the terminal velocity u is attained:

$$F = k_1 u^2 d_p^2$$

since the area is proportional to d_p^2 and the accelerating force $= (\rho_s - \rho)gk_2 d_p^3$ where k_2 is a constant depending on the shape of the particle and d_p is a mean projected area.

When the terminal velocity is reached,

$$k_1 u^2 d_p^2 = (\rho_s - \rho)gk_2 d_p^3$$

$$\therefore \qquad u = [(\rho_s - \rho)gk_3 d_p]^{0\cdot5}$$

In order to achieve separation, the terminal velocity of the smallest particle (diameter d_1) of the dense material must be at least equal to that of the largest particle (diameter d_2) of the light material. That is:

$$[(9800 - 1000)9\cdot81k_3 d_1]^{0\cdot5} = [(3700 - 1000)9\cdot81k_3 d_2]^{0\cdot5}$$

$$\therefore \qquad (d_2/d_1) = 8800/2700$$

$$\therefore \qquad = 3\cdot26$$

which is the maximum range of sizes which can be separated if the terminal velocities are attained.

If the particles are allowed to settle in the fluid for only a very short time, they will not attain their terminal falling velocities and a better degree of separation can be obtained. A particle of the denser material will have an initial acceleration $g(1 - \rho/\rho_s)$ because there is no fluid friction when the relative velocity is zero. Thus the initial velocity is a function of density only and is unaffected by size and shape. A very small particle of the denser material will therefore always commence settling at a greater rate than a large particle of the less dense material. Theoretically it should be possible to separate materials completely irrespective of the size range provided that the periods of settling are sufficiently short. In practice the required periods will often be so short that it is impossible to make use of this principle alone. As the time of settling increases the larger particles of the less dense material catch up and overtake the smaller particles of the denser material.

If the total drag force is proportional to the velocity squared, i.e. $\dot{y}^2$, then the equation of motion for a particle falling downwards under the influence of gravity may be written as:

$$m\ddot{y} = mg(1 - \rho/\rho_s) - k_1 \dot{y}^2$$

$$\therefore \qquad \ddot{y} = g(1 - \rho/\rho_s) - (k_1/m) \dot{y}^2$$

or $$\ddot{y} = b - c\dot{y}^2$$

where $b = g(1 - \rho/\rho_s)$, $c = k_1/m$, and k_1 is a proportionality constant.

$$\therefore \qquad d\dot{y}/(b - c\dot{y}^2) = dt$$

or $$d\dot{y}/(f^2 - \dot{y}^2) = c\, dt$$

where $f = \sqrt{(b/c)}$.

Integrating, $(1/2f) \ln[(f + \dot{y})/(f - \dot{y})] = ct + k_4$

When $t = 0$, $\qquad\qquad \dot{y} = 0$ and $k_4 = 0$

$$\therefore \qquad (1/2f) \ln[(f + \dot{y})/(f - \dot{y})] = ct$$

$$(f + \dot{y})/(f - \dot{y}) = e^{2fct}$$

$$f - \dot{y} = 2f/(1 + e^{2fct})$$

$$y = ft - 2f \int dt/(1 - e^{2fct})$$

$$y = ft - (1/c)\ln\left[e^{2fct}/(1 + e^{2fct})\right] + k_5$$

When $t = 0$, $y = 0$ and $k_5 = (1/c)\ln 0.5$

$$\therefore \qquad \underline{y = ft - (1/c)\ln(0.5\,e^{2fct})/(1 + e^{2fct})}$$

where $f = \sqrt{(b/c)}$, $b = g(1 - \rho/\rho_s)$, and $c = k_1/m$.

Problem 3.13

Salt, of specific gravity 2·35, is charged to the top of a reactor containing a 3 m depth of aqueous liquid (specific gravity 1·1 and viscosity $2\,\text{mN s/m}^2$) and the crystals must dissolve completely before reaching the bottom. If the rate of dissolution of the crystals is given by the relation:

$$-dd/dt = 3 \times 10^{-4} + 2 \times 10^{-4}u$$

where d is the size of the crystal (cm) at time t (s) and u its velocity in the fluid (cm/s); calculate the maximum size of crystal which can be charged. The inertia of the particles can be neglected and the resistance force can be taken as given by Stokes' law ($3\pi\mu du$), d being taken as the equivalent spherical diameter of the particle.

Solution

Assuming the salt always travels at its terminal velocity, then for the Stokes' law region, this is given by equation 3.17:

$$u_0 = (d^2 g/18\mu)(\rho_s - \rho)$$

or, in this case,

$$u_0 = (d^2 \times 9.81/18 \times 2 \times 10^{-3})(2350 - 1100)$$

$$= 3.46 \times 10^5 d^2 \text{ m/s}$$

The rate of dissolution,

$$-dd/dt = 3 \times 10^{-6} + 2 \times 10^{-4}u\,\text{m/s}$$

and substituting,

$$dd/dt = -3 \times 10^{-6} - (2 \times 10^{-4} \times 3.406 \times 10^5 d^2)$$

$$= -3 \times 10^{-6} - 68.1d^2$$

The velocity at any point h from the top of the reactor is $u = dh/dt$,

$$\frac{dh}{dd} = \frac{dh}{dt}\frac{dt}{dd} = 3.406 \times 10^5 d^2/(-3 \times 10^{-6} - 68.1d^2)$$

and
$$\int_0^3 dh = -\int_D^0 \frac{3\cdot406 \times 10^5 d^2\, dd}{3 \times 10^{-6} + 68\cdot1 d^2}$$

$\therefore$
$$3 = 3\cdot406 \times 10^5 \left(\int_0^D \frac{dd}{B} - \frac{A}{B^2} \int_0^D \frac{dd}{(A/B) + d^2} \right)$$

where $A = 3 \times 10^{-6}$ and $B = 68\cdot1$.

$\therefore$
$$3 = 3\cdot406 \times 10^5 \left\langle \left[\frac{d}{B} \right]_0^D - \left[\frac{A}{B^2} \frac{1}{\sqrt{(A/B)}} \tan^{-1}\left(\frac{d}{\sqrt{(A/B)}} \right) \right]_0^D \right\rangle$$

$\therefore$
$$3 = (3\cdot406 \times 10^5/B)[D - (A/B)^{\frac{1}{2}} \tan^{-1} D(A/B)^{-\frac{1}{2}}]$$

Substituting for A and B,

$$D = 6 \times 10^{-4} + 2\cdot1 \times 10^{-4} \tan^{-1}(4\cdot76 \times 10^3 D)$$

and solving by trial and error,

$$D = 8\cdot8 \times 10^{-4}\,\mathrm{m} \quad \text{or} \quad \underline{0\cdot88\,\mathrm{mm}}$$

The integration may also be carried out numerically and the working is as follows:

d	d^2	$\left(\dfrac{3\cdot406 \times 10^5 d^2}{3 \times 10^{-6} + 68\cdot1 d^2}\right)$	Interval of d	Mean value of function in interval	Integral in interval	Total integral
0	0	0				
			1×10^{-4}	$4\cdot63 \times 10^2$	0·0463	0·0463
1×10^{-4}	1×10^{-8}	$9\cdot25 \times 10^2$				
			1×10^{-4}	$1\cdot65 \times 10^3$	0·1653	0·2116
2×10^{-4}	4×10^{-8}	$2\cdot38 \times 10^3$				
			1×10^{-4}	$2\cdot86 \times 10^3$	0·2869	0·4985
3×10^{-4}	9×10^{-8}	$3\cdot358 \times 10^3$				
			1×10^{-4}	$3\cdot64 \times 10^3$	0·364	0·8625
4×10^{-4}	$1\cdot6 \times 10^{-7}$	$3\cdot922 \times 10^3$				
			1×10^{-4}	$4\cdot09 \times 10^3$	0·409	1·2715
5×10^{-4}	$2\cdot5 \times 10^{-7}$	$4\cdot25 \times 10^3$				
			1×10^{-4}	$4\cdot35 \times 10^3$	0·435	1·706
6×10^{-4}	$3\cdot6 \times 10^{-7}$	$4\cdot46 \times 10^3$				
			1×10^{-4}	$4\cdot52 \times 10^3$	0·452	2·158
7×10^{-4}	$4\cdot9 \times 10^{-7}$	$4\cdot589 \times 10^3$				
			1×10^{-4}	$4\cdot634 \times 10^3$	0·463	2·621
8×10^{-4}	$6\cdot4 \times 10^{-7}$	$4\cdot679 \times 10^3$				
			1×10^{-4}	$4\cdot709 \times 10^3$	0·471	3·09
9×10^{-4}	$8\cdot1 \times 10^{-7}$	$4\cdot74 \times 10^3$				

From which $D = \underline{0\cdot9\,\mathrm{mm}}$.

The acceleration of the particle to its terminal velocity has been neglected, and in practice the time to reach the bottom of the reactor would be slightly longer, allowing a larger crystal to dissolve completely.

Problem 3.14

A balloon weighing 7 g is charged with hydrogen to a pressure of $104 \, \text{kN/m}^2$. The balloon is released from ground level and, as it rises, hydrogen escapes in order to maintain a constant differential pressure of $2 \cdot 7 \, \text{kN/m}^2$ under which condition the diameter of the balloon is $0 \cdot 3 \, \text{m}$. If conditions are assumed to remain isothermal at 273 K as the balloon rises, what is the ultimate height reached and how long does it take to rise through the first 3000 m?

It may be assumed that the value of the Reynolds number with respect to the balloon exceeds 500 throughout, so that the resistance coefficient is constant at $0 \cdot 22$. Neglect the inertia of the balloon, i.e. assume that it is rising at its equilibrium velocity at any moment.

Solution

Volume of balloon $= (4/3) \, \pi (0 \cdot 15)^3 = 0 \cdot 0142 \, \text{m}^3$.
Mass of balloon $= 7 \, \text{g}$ or $0 \cdot 007 \, \text{kg}$.
The upthrust $=$ weight of air at $p \, \text{N/m}^2 -$ weight of hydrogen at $(p + 2700) \, \text{N/m}^2$.
If ρ_a is the density of air at $101{,}300 \, \text{N/m}^2$ and 273 K $(28 \cdot 9/22 \cdot 4) = 1 \cdot 29 \, \text{kg/m}^3$, where the mean molecular weight of air is taken as $28 \cdot 9 \, \text{kg/mol}$, then the net upthrust force W is given by:

$$W = 9 \cdot 81 \, \{0 \cdot 0142 \, [(\rho_a p/101{,}300) - \rho_a(2/28 \cdot 9)(p + 2700)/101{,}300] - 0 \cdot 007\}$$

$$= 0 \cdot 139 \, [0 \cdot 000 \, 012 \, 7p - 0 \cdot 000 \, 000 \, 881 \, (p + 2700)] - 0 \cdot 0687$$

$$= 0 \cdot 000 \, 001 \, 64p - 0 \cdot 0690 \, \text{N}$$

The balloon will stop when $W = 0$, that is when

$$p = 0 \cdot 0690/0 \cdot 000 \, 001 \, 64 = 42{,}092 \, \text{N/m}^2$$

The variation of pressure with height is given by:

$$g \, dz + v \, dp = 0$$

But $v = (1/\rho_a)(101{,}300/p) \, \text{m}^3$ for isothermal conditions

$\therefore$ $dz + [101{,}300/(9 \cdot 81 \times 1 \cdot 29p)] \, dp = 0$

$\therefore$ $z_2 - z_1 = 8005 \ln (101{,}300/p)$

When $p = 42{,}092 \, \text{N/m}^2$,

$$z_2 - z_1 = 8005 \ln (101{,}300/42{,}092)$$

$$= \underline{7030 \, \text{m}}$$

The resistance force R on the balloon is given by:

$$(R/\rho_a u^2) = 0 \cdot 22$$

or $R = 0 \cdot 22 \rho_a (p/101{,}300)(\pi \times 0 \cdot 3^2/4)(dz/dt)^2 \, \text{N/m}^2$

$$= 1 \cdot 98 \times 10^{-7} p(dz/dt)^2$$

This must be equal to the net upthrust force W, or:

$$0.000\,001\,64p - 0.0690 = 1.98 \times 10^{-7} p\,(dz/dt)^2$$

$$\therefore \qquad (dz/dt)^2 = 8.28 - 3.49 \times 10^5/p$$

But $$z = 8005 \ln(101,300/p)$$

$$\therefore \qquad (dz/dt)^2 = 8.28 - (3.49 \times 10^5 \, e^{z/8005}/101,300)$$

$$\therefore \qquad (dz/dt) = 1.89(2.41 - e^{1.25 \times 10^{-4}z})^{0.5}$$

The time taken to rise 3000 m is therefore given by:

$$t = (1/1.89) \int_0^{3000} dz/(2.41 - e^{1.25 \times 10^{-4}z})^{0.5}$$

Writing the integral as $$I = \int_0^{3000} dz/(a - e^{bz})^{0.5}$$

and putting $$(a - e^{bz}) = x^2$$

$$\therefore \qquad dz = 2x\,dx/[b(a - x^2)]$$

and $$I = (-2/b) \int dx/(a - x^2)$$

$$= (-2/b)(1/2\sqrt{a})\left[\ln \frac{\sqrt{a} - \sqrt{(a - e^{bz})}}{\sqrt{a} + \sqrt{(a - e^{bz})}}\right]_0^{3000}$$

$$= (1/b\sqrt{a})\ln \frac{[\sqrt{a} - \sqrt{(a - e^{3000b})}][\sqrt{a} + \sqrt{(a - 1)}]}{[\sqrt{a} + \sqrt{(a - e^{3000b})}][\sqrt{a} - \sqrt{(a - 1)}]}$$

Now, $$a = 2.41 \quad\text{and}\quad b = 1.25 \times 10^{-4}$$

$$\therefore \qquad I = 5161 \ln[(1.55 - 0.977)/(1.55 + 0.977)][(1.55 - 1.19)/(1.55 + 1.19)]$$

$$= 2816$$

$$\therefore \qquad t = [2816(1/1.89)]$$

$$= \underline{\underline{1490\,s \quad (25\,min)}}$$

Problem 3.15

A mixture of quartz of specific gravity 3·7 and galena of specific gravity 9·8 whose size range is 0·3 to 1 mm is to be separated by a sedimentation process. If Stokes' law is assumed to be applicable, what is the minimum density required for the liquid if the particles all settle at their terminal velocities?

Consideration was given to devising a separating system using water as the liquid. In this case the particles were to be allowed to settle for a series of short-time intervals so that the smallest particle of galena settled a larger distance than the largest particle of quartz. What approximately is the maximum permissible settling period?

According to Stokes' law the resistance force F acting on a particle of diameter d settling at a velocity u in a fluid of viscosity μ is given by:

$$F = 3\pi\mu du$$

Viscosity of water $= 1 \text{ mN s/m}^2$.

Solution

For streamline conditions, equation 1.31 applies:

$$d_B/d_A = [(\rho_A - \rho)/(\rho_B - \rho)]^{0\cdot5}$$

For separation it is necessary that a large particle of the less dense material does not overtake a small particle of the dense material, i.e.

$$(1/0\cdot3) = [(9800 - \rho)/(3700 - \rho)]^{0\cdot5}$$

$$\therefore \qquad \rho = \underline{\underline{3097 \text{ kg/m}^3}}$$

Assuming Stokes' law is valid, the distance travelled including the period of acceleration is given by equation 3.53:

$$y = (b/a)t + (v/a) - (b/a^2) - [(b/a^2) - (v/a)]e^{-at}$$

When the initial velocity $v = 0$,

$$y = (b/a)t + (b/a^2)(e^{-at} - 1)$$

where $b = g(1 - \rho/\rho_s)$ (equation 3.52)

and $a = 18\mu/d^2\rho_s$

For a small particle of galena:

$b = 9\cdot81(1 - 1000/9800) = 8\cdot81 \text{ m/s}^2$

$a = (18 \times 1 \times 10^{-3})/[(0\cdot3 \times 10^{-3})^2 \times 9800] = 20\cdot4/\text{s}$

For a large particle of quartz:

$b = 9\cdot81(1 - 1000/3700) = 7\cdot15 \text{ m/s}^2$

$a = (18 \times 1 \times 10^{-3})/[(1 \times 10^{-3})^2 \times 3700] = 4\cdot86/\text{s}$

In order to achieve separation, these particles must travel at least the same distance in time t or:

$$(8\cdot81/20\cdot4)t + (8\cdot81/20\cdot4^2)(e^{-20\cdot4t} - 1) = (7\cdot15/4\cdot86)t + (7\cdot15/4\cdot86^2)(e^{-4\cdot86t} - 1)$$

$$\therefore \qquad (0\cdot0212\,e^{-20\cdot4t} - 0\cdot303\,e^{-4\cdot86t}) = 1\cdot039t - 0\cdot282$$

Solving by trial and error, $t = \underline{\underline{0\cdot05 \text{ s}}}$

FLOW OF FLUIDS THROUGH GRANULAR BEDS AND PACKED COLUMNS

Problem 4.1

In a contact sulphuric acid plant the secondary converter is a tray type converter, 2·3 m in diameter with the catalyst arranged in three layers, each 0·45 m thick. The catalyst is in the form of cylindrical pellets 9·5 mm in diameter and 9·5 mm long. The void fraction is 0·35. The gas enters the converter at 675 K and leaves at 720 K. Its inlet composition is

$$SO_3 \ 6\cdot6, \quad SO_2 \ 1\cdot7, \quad O_2 \ 10\cdot0, \quad N_2 \ 81\cdot7 \ \text{mol} \%$$

and its exit composition

$$SO_3 \ 8\cdot2, \quad SO_2 \ 0\cdot2, \quad O_2 \ 9\cdot3, \quad N_2 \ 82\cdot3 \ \text{mol} \%$$

The gas flow rate is 0·68 kg/m²s. Calculate the pressure drop through the converter.

$$\mu = 0\cdot032 \ \text{mN s/m}^2$$

Solution

This problem will be solved by three different methods.

(a) *Chilton and Colburn*

$$Re' = \rho u d / \mu \quad \text{and} \quad \frac{(-\Delta P) \ d}{2\rho u^2 \ l} = \phi_1'$$

For $Re' < 40$: $\qquad \phi_1' = 850/Re'$

For $Re' > 40$: $\qquad \phi_1' = 38/Re'^{0\cdot15}$

The mean molecular weight

$$= (0\cdot066 \times 80) + (0\cdot017 \times 64) + (0\cdot1 \times 32) + (0\cdot817 \times 28)$$

$$= 32\cdot44 \ \text{kg/kmol} \quad \text{at the inlet,}$$

and at the outlet:

$$(0\cdot082 \times 80) + (0\cdot002 \times 64) + (0\cdot093 \times 32) + (0\cdot823 \times 28) = 32\cdot71 \ \text{kg/kmol}$$

Inlet temperature = 675 K.
Outlet temperature = 720 K.

$\therefore$ average temperature = 697·5 K.

Average molecular weight = 32·58 kg/kmol.

$$\text{Average gas density} = \frac{32\cdot58}{22\cdot4} \times \frac{273}{697\cdot5} = 0\cdot569\,\text{kg/m}^3.$$

Reynolds number = $\rho u d/\mu = Gd/\mu$

$$= 0\cdot68 \times 9\cdot5 \times 10^{-3}/0\cdot032 \times 10^{-3}$$

$$= 202$$

$$-\Delta P = \frac{2\rho u^2 l}{d} \times \frac{38}{(202)^{0\cdot15}}$$

Average gas velocity $u = 0\cdot68/0\cdot569 = 1\cdot20\,\text{m/s}$

$$\therefore \qquad -\Delta P = \frac{2 \times 0\cdot569 \times (1\cdot20)^2 \times (3 \times 0\cdot45)}{9\cdot5 \times 10^{-3}} \times \frac{38}{(202)^{0\cdot15}}$$

$$= 3\cdot99 \times 10^3\,\text{N/m}^2$$

$$= 4\cdot0\,\text{kN/m}^2$$

(b) *Rose*

Rose defined Re' in the same way as Chilton and Colburn but instead of ϕ_1' used ϕ_1, i.e. $-\Delta Pd/\rho u^2 l$. Defining d as the diameter of a sphere with the same specific surface as the material forming the bed, his correlation was presented as:

$$\phi_1 = 1000/Re' + 125/Re^{0\cdot5} + 14$$

For cylindrical pellets of length = diameter = d:

$$S = \left(2\frac{\pi}{4}d^2 + \pi d^2\right)\bigg/\frac{\pi}{4}d^3 = 6/d$$

which is the same as for spheres.

$$\therefore \qquad d = 9\cdot5\,\text{mm} \quad \text{and} \quad Re' = 202 \quad \text{as before.}$$

Then $\qquad\qquad \phi_1 = 1000/202 + 125/202 + 14$

$$= 27\cdot7$$

$$\therefore \qquad \Delta P = 27\cdot7 \times 0\cdot569 \times (1\cdot20)^2 \times (3 \times 0\cdot45)/9\cdot5 \times 10^{-3}$$

$$= 3\cdot23 \times 10^3\,\text{N/m}^2$$

$$= 3\cdot2\,\text{kN/m}^2$$

(c) *Carman*

$$\frac{R}{\rho u^2} = \frac{e^3}{5(1-e)}\frac{\Delta P}{l}\frac{1}{\rho u^2} \qquad\qquad \text{(equation 4.13)}$$

$$\frac{R}{\rho u^2} = 5/Re_1 + 0\cdot4/Re_1^{0\cdot1} \qquad \text{(equation 4.14)}$$

$$Re_1 = \frac{G}{5(1-e)\,\mu} \qquad \text{(equation 4.11)}$$

$$S = 6/d = 6/9\cdot5 \times 10^{-3} = 631\,\text{m}^2/\text{m}^3$$

$$\therefore \qquad Re_1 = 0\cdot68/631 \times 0\cdot65 \times 0\cdot032 \times 10^{-3} = 51\cdot8$$

$$\therefore \qquad \frac{R}{\rho u^2} = \frac{5}{51\cdot8} + \frac{0\cdot4}{(51\cdot8)^{0\cdot1}} = 0\cdot366$$

From equation 4.13:

$$\Delta P = 0\cdot366 \times 631 \times 0\cdot65 \times (3 \times 0\cdot45) \times 0\cdot569 \times (1\cdot20)^2/(0\cdot35)^3$$
$$= 3\cdot87 \times 10^3\,\text{N/m}^2$$
$$= \underline{\underline{3\cdot9\,\text{kN/m}^2}}$$

Problem 4.2

Show how an equation for the pressure drop in a packed column can be modified to apply to cases where the total pressure and the pressure drop are of the same order of magnitude.

Two heat-sensitive organic liquids (average molecular weight = 155 kg/kmol) are to be separated by vacuum distillation in a 100 mm diameter column packed with 6 mm stoneware Raschig rings. The number of theoretical plates required is 16 and it has been found that the HETP is 150 mm. If the product rate is 5 g/s at a reflux ratio of 8, calculate the pressure in the condenser so that the temperature in the still does not exceed 395 K (equivalent to a pressure of 8 kN/m²). Assume $a = 800\ \text{m}^2/\text{m}^3$, $\mu = 0\cdot02\ \text{mN s/m}^2$, $e = 0\cdot72$, and neglect the temperature changes and the correction for liquid flow.

Solution

The modified Reynolds number Re_1 is defined by:

$$Re_1 = \frac{\mu\rho}{S(1-e)\mu} = \frac{G}{S(1-e)\mu} \qquad \text{(equation 4.11)}$$

Ergun's equation (4.18) may be rewritten as:

$$\frac{R}{\rho u^2} = 4\cdot17/Re_1 + 0\cdot29 \qquad \text{(equation 4.19)}$$

Hence

$$\frac{R}{\rho u^2} = \frac{4\cdot17 S(1-e)\mu}{G} + 0\cdot29$$

Equation 4.13 states:

$$\frac{R}{\rho u^{2\bullet}} = \frac{e^3}{S(1-e)}\frac{(-dP)}{dl}\frac{1}{\rho u^2}$$

$$= \frac{e^3}{S(1-e)}\frac{(-dP)}{dl}\frac{\rho}{G^2}$$

$$-\rho\frac{dP}{dl} = \frac{R}{\rho u^2}\frac{S(1-e)}{e^3}G^2$$

$$-\int \rho\,dP = \frac{R}{\rho u^2}\frac{S(1-e)}{e^3}G^2\int dl$$

$$= \frac{R}{\rho u^2}\frac{S(1-e)}{e^3}G^2 l$$

In this problem, $a = 800\,\text{m}^2/\text{m}^3 = S(1-e)$

Product rate $= 0.5\,\text{g/s}$ and if the reflux ratio $= 8$, then:
Vapour rate $= 4.5\,\text{g/s}$
and $G = 4.5 \times 10^{-3}/(\pi/4)(0.1)^2 = 0.573\,\text{kg/m}^2\,\text{s}$
          $\mu = 0.02 \times 10^{-3}\,\text{N s/m}^2$
          $e = 0.72$
$\therefore$ $Re_1 = 0.573/800 \times 0.28 \times 0.02 \times 10^{-3} = 128$

$$\frac{R}{\rho u^2} = 4.17/128 + 0.29 = 0.32$$

Hence $-\int \rho\,dP = 0.32 \times 800 \times 0.28 \times (0.573)^2 \times 2.4/(0.72)^3$

since $l = 16 \times 0.15 = 2.4\,\text{m}$

$$-\int \rho\,dP = 151.3$$

Now $\rho = \rho_s \times P/P_s$

where s refers to the still.
The vapour density in the still

$$\rho_s = \frac{155}{22.4} \times \frac{273}{395} \times \frac{P_s}{101.3 \times 10^3} = 4.73 \times 10^{-5}P_s$$

$\therefore$ $\rho = 4.73 \times 10^{-5}P$

$\therefore$ $-\int_{P_c}^{P_s} \rho\,dP = -\int_{P_c}^{P_s} 4.73 \times 10^{-5}P\,dP = 4.73 \times 10^{-5}(P_s^2 - P_c^2)$

where $P_c = $ condenser pressure.

Now $(P_s - P_c) = -\Delta P$ and if $P_s \simeq -\Delta P$:

$$(P_s^2 - P_c^2) \simeq \Delta P^2$$

$$\therefore \qquad \Delta P^2 = 151 \cdot 3/(4 \cdot 73 \times 10^{-5})$$

$$\Delta P = 1790 \, \text{N/m}^2$$

Now
$$P_s = 8000 \, \text{N/m}^2$$

$$\therefore \qquad P_c = 6210 \, \text{N/m}^2 = \underline{\underline{6 \cdot 2 \, \text{kN/m}^2}}$$

Problem 4.3

A column 0·6 m diameter and 4 m tall, packed with 25 mm ceramic Raschig rings, is used in a gas absorption process carried out at atmospheric pressure and 293 K. If the liquid and gas can be considered to have the properties of water and air, and their flow rates are 6·5 and 0·6 kg/m² s respectively, what will be the pressure drop across the column?

Use (a) Carman's method, (b) one other method, and compare the results obtained. How much can the liquid rate be increased before the column will flood?

Solution

(a) *Carman's method*

Equation 4.17 presents Carman's correlation for flow through randomly packed beds as:

$$R/\rho u^2 = 5/Re_1 + 1 \cdot 0/Re_1^{0 \cdot 1} \qquad \text{(equation 4.17)}$$

where
$$R/\rho u^2 = \frac{e^3}{S(1-e)} \frac{\Delta P}{l} \frac{1}{\rho u^2} \qquad \text{(equation 4.13)}$$

and
$$Re_1 = \frac{G}{S(1-e)\mu} \qquad \text{(equation 4.11)}$$

Using the data given in the problem:

$$\rho_{\text{air}} = \frac{29}{22 \cdot 4} \times \frac{273}{293} = 1 \cdot 21 \, \text{kg/m}^3$$

$$G = 0 \cdot 6 \, \text{kg/m}^2 \, \text{s}$$

$$\therefore \qquad u = 0 \cdot 6/1 \cdot 21 = 0 \cdot 496 \, \text{m/s}$$

From Problem 4.5 for 25 mm Raschig rings,

$$S = 656 \, \text{m}^2/\text{m}^3 \quad \text{and} \quad e = 0 \cdot 71$$

$$\therefore \qquad Re_1 = 0 \cdot 6/656 \times 0 \cdot 29 \times 0 \cdot 018 \times 10^{-3} = 175$$

$$\frac{R}{\rho u^2} = \frac{(0 \cdot 71)^3}{656 \times 0 \cdot 29} \times \frac{(-\Delta P)}{4} \times \frac{1}{1 \cdot 21 \times (0 \cdot 496)^2} = 1 \cdot 71 \times 10^{-3} P$$

$$\therefore \qquad 1 \cdot 71 \times 10^{-3}(-\Delta P) = 5/175 + 1 \cdot 0/(175)^{0 \cdot 1} = 0 \cdot 625$$

and $\Delta P = 365 \, \text{N/m}^2$

Figure 4.23 may be used to allow for the effect of liquid flow. The correction factor is found from Fig. 4.23 to be 1·8 so that:

$$P = 365 \times 1\cdot8 = 670 \, \text{N/m}^2$$

(b) *Morris and Jackson's method*

The wetting rate $L_w = u_L/S_B = 6\cdot5/1000 \times 190 = 3\cdot42 \times 10^{-5} \, \text{m}^3/\text{s m}^2$

since $S_B = S(1 - e) = 190 \, \text{m}^2/\text{m}^3$

From Fig. 4.25 the number of velocity heads lost $= N = 1050$.

Then $\Delta P = 0\cdot5N\rho_G u_G^2 l$ (equation 4.36)

$$= 0\cdot5 \times 1050 \times 1\cdot21 \times 0\cdot496^2 \times 4$$

$$= 625 \, \text{N/m}^2$$

(c) Figure 4.28 may be used to find at what liquid flow rate the column will flood.

$$\frac{u_G^2 S_B \, \rho_G}{ge^3 \, \rho_L}\left(\frac{\mu_L}{\mu_w}\right)^{0\cdot2} = \frac{0\cdot496^2 \times 190 \times 1\cdot21}{9\cdot81 \times (0\cdot71)^3 \times 1000} = 0\cdot0161$$

From the flooding curve, $\dfrac{L}{G}\sqrt{\dfrac{\rho_G}{\rho_L}} = 1\cdot2$

and $L = 1\cdot2 \times 0\cdot6\sqrt{(1000/1\cdot21)}$

$$= 20\cdot7 \, \text{kg/m}^2 \, \text{s}$$

Problem 4.4

A packed column, 1·2 m in diameter and 9 m tall, and packed with 25 mm Raschig rings, is used for the vacuum distillation of a mixture of isomers of molecular weight 155 kg/kmol. The mean temperature is 373 K, the pressure at the top of the column is maintained at 0·13 kN/m², and the still pressure ranges between 1·3 and 3·3 kN/m². Obtain an expression for the pressure drop on the assumption that it is not appreciably affected by the liquid flow and can be calculated using a modified form of Carman's equation (4.17). Show that, over the range of operating pressures used, the pressure drop is approximately directly proportional to the mass rate of flow of vapour, and calculate the pressure drop at a vapour rate of 0·125 kg/m² s.

Data: Specific surface of packing $S = 650 \, \text{m}^2/\text{m}^3$.

Mean voidage of bed $e = 0\cdot71$.

Viscosity of vapour $= 0\cdot018 \, \text{mN s/m}^2$.

Molecular volume $= 22\cdot4 \, \text{m}^3/\text{kmol}$.

Solution

The proof that the pressure drop is approximately proportional to the mass flow rate of vapour is illustrated in Problem 4.5. Using the data specified in this problem:

$$Re_1 = G/S(1-e)\mu$$

$$= 0{\cdot}125/660 \times 0{\cdot}29 \times 0{\cdot}018 \times 10^{-3} = 36{\cdot}3$$

The modified Carman's equation states:

$$R/\rho u^2 = 5/Re_1 + 1/Re_1^{0{\cdot}1}$$

$$= 5/36{\cdot}3 + 1/(36{\cdot}3)^{0{\cdot}1} = 0{\cdot}836$$

As in Problem 4.2:

$$\frac{R}{\rho u^2} = \frac{e^3}{S(1-e)}\frac{(-dP)}{dl}\frac{1}{\rho u^2} \qquad \text{(equation 4.13)}$$

$$= \frac{e^3}{S(1-e)}\frac{(-dP)}{dl}\frac{\rho}{G^2}$$

$$\therefore \quad -\int \rho \, dP = \frac{R}{\rho u^2}\frac{S(1-e)}{e^3}G^2l$$

$$= 0{\cdot}836 \times 650 \times 0{\cdot}29 \times 9G^2/(0{\cdot}71)^3$$

$$= 3690G^2$$

Now $\rho/P = \rho_s/P_s$ where subscript s refers to the still,

$$\rho_s = \frac{155}{22{\cdot}4} \times \frac{273}{373} \times \frac{P_s}{101{\cdot}3 \times 10^3} = 5 \times 10^{-5}P_s$$

$$\rho_s/P_s = 5 \times 10^{-5}$$

$$\therefore \quad \rho = 5 \times 10^{-5}P$$

$$-\int_{P_c}^{P_s} \rho \, dP = 2{\cdot}5 \times 10^{-5}(P_s^2 - P_c^2)$$

Now $P_s - P_c = -\Delta P$, and if $\Delta P \simeq -P_s$, $\quad (P_s^2 - P_c^2) \simeq \Delta P^2$

$$-\int_{P_c}^{P_s} \rho \, dP = 2{\cdot}5 \times 10^{-5}\Delta P^2 = 3690G^2$$

i.e. $$\Delta P \simeq G$$

If $G = 0{\cdot}125$, $-\Delta P = [3690 \times (0{\cdot}125)^2/2{\cdot}5 \times 10^{-5}]^{0{\cdot}5}$

$$= 1520 \, \text{N/m}^2$$

$$= 1{\cdot}52 \, \text{kN/m}^2$$

Problem 4.5

A packed column, 1·22 m in diameter and 9 m tall, and packed with 25 mm Raschig rings, is used for the vacuum distillation of a mixture of isomers of molecular weight 155 kg/kmol. The mean temperature is 373 K, the pressure at the top of the column is maintained at 0·13 kN/m², and the still pressure is 1·3 kN/m². Obtain an expression for the pressure drop on the assumption that it is not appreciably affected by the liquid flow and can be calculated using the modified form of Carman's equation.

Show that, over the range of operating pressures used, the pressure drop is approximately directly proportional to the mass rate of flow of vapour, and calculate approximately the flow rate of vapour.

Data: Specific surface of packing $S = 656 \, \text{m}^2/\text{m}^3$.

Mean voidage of bed $e = 0.71$.

Viscosity of vapour $\mu = 0.018 \, \text{mN s/m}^2$.

Kilogram molecular volume $= 22.4 \, \text{m}^3/\text{kmol}$.

Solution

The modified form of Carman's equation states:

$$R/\rho u^2 = 5/Re_1 + (1/Re_1)^{0.1} \qquad \text{(equation 4.17)}$$

where
$$Re_1 = G/S(1-e)\mu$$

$$Re_1 = \frac{G}{656(1-0.71) \times 0.018 \times 10^{-3}} = 292G$$

$$\therefore \quad \frac{R}{\rho u^2} = \frac{5}{292G} + \left(\frac{1}{292G}\right)^{0.1} = \frac{0.017}{G} + \frac{0.57}{G^{0.1}}$$

Now
$$\frac{R}{\rho u^2} = \frac{e^3}{S(1-e)} \frac{(-dP)}{dl} \frac{\rho}{G^2}$$

and as in the previous problem, by substitution:

$$-\int \rho \, dP = 81.4G + 2730G^{1.9}$$

As before,
$$-\int \rho \, dP = 2.5 \times 10^{-5} \Delta P^2$$

$$\therefore \quad \Delta P^2 = 3.26 \times 10^6 G + 1.09 \times 10^8 G^{1.9}$$

If the first term is neglected:

$$-\Delta P = 1.05 \times 10^4 G^{0.95}$$

If $-\Delta P = 1300 - 130 = 1170 \, \text{N/m}^2$:

$$G = 0.099 \, \text{kg/m}^2 \, \text{s}$$

SECTION 5

SEDIMENTATION

Problem 5.1

A slurry containing 5 kg of water per kg of solids is to be thickened to a sludge containing 1·5 kg of water per kg of solids in a continuous operation. Laboratory tests using five different concentrations of the slurry yielded the following results:

Concentration (kg water/kg solid):	5·0	4·2	3·7	3·1	2·5
Rate of sedimentation (mm/s):	0·17	0·10	0·08	0·06	0·042

Calculate the minimum area of a thickener to effect the separation of 0·6 kg of solids per second.

Solution

Basis: 1 kg of solids: 1·5 kg water is carried away in underflow, balance in overflow, $V = 1·5$.

Concentration U	Water to overflow $(U - V)$	Sedimentation rate u_c (mm/s)	$(U - V)/u_c$ (s/mm)
5·0	3·5	0·17	20·56
4·2	2·7	0·10	27·0
3·7	2·2	0·08	27·5
3·1	1·6	0·06	26·67
2·5	1·0	0·042	23·81

Maximum value of $(U - V)/u_c = 27·5$ s/mm or 27,500 s/m.
From equation 5.101:

$$A = [(U - V)/u_c](W/\rho)$$

$$= 27,500(0·6/1000)$$

$$= \underline{\underline{16·5\,\mathrm{m}^2}}$$

Problem 5.2

If a centrifuge is 0·9 m diameter and rotates at 20 Hz, at what speed should a laboratory centrifuge of 150 mm diameter run if it is to duplicate plant conditions?

Solution

If a particle of mass m is rotating at radius x with an angular velocity w, it is subjected to a centrifugal force $m \times w^2$ in a radial direction and a gravitational force mg in a vertical direction. The ratio of the centrifugal to gravitational forces, xw^2/g, is a measure of the separating power of the machine, and for duplicate conditions this must be the same in both machines.

In this case:
$$x_1 = 0.45 \, \text{m}$$
$$w_1 = 20 \times 2\pi = 40\pi \, \text{rad/s}$$
$$x_2 = 0.075 \, \text{m}$$
$$\therefore \quad 0.45(40\pi)^2/g = 0.075 w_2^2/g$$
$$w_2 = \sqrt{[6(40\pi)^2]}$$
$$= 2.45 \times 40\pi = 98\pi \, \text{rad/s}$$

and the speed of rotation
$$= 98\pi/2\pi$$
$$= \underline{\underline{49 \, \text{Hz}}}$$

Problem 5.3

What is the maximum safe speed of rotation of a phosphor-bronze centrifuge basket, 0.3 m diameter and 5 mm thick, when it contains a liquid of density $1000 \, \text{kg/m}^3$ forming a layer 75 mm thick at the walls? Take the density of phosphor-bronze as $8900 \, \text{kg/m}^3$ and the safe working stress as $55 \, \text{MN/m}^2$.

Solution

The centrifugal pressure due to the liquid is given by (the nomenclature is defined in Problem 5.4):
$$P_c = 0.5\rho w^2 (b^2 - x^2)$$
$$= 0.5 \times 1000 \times w^2 (0.15^2 - 0.075^2)$$
$$= 8.438 w^2 \, \text{N/m}^2$$

The stress in the walls of the basket is given by:
$$f = (b/\delta)(P_c + \rho_m \, \delta b w^2)$$
$$= (0.15/0.005)(8.438 w^2 + 8900 \times 0.005 \times 0.15 w^2)$$
$$= 453 w^2 \, \text{N/m}^2$$

The maximum speed of rotation is therefore
$$w = \sqrt{(55 \times 10^6/453)}$$
$$= 348 \, \text{rad/s} \quad \text{or} \quad \underline{\underline{55.5 \, \text{Hz}}} \quad (3327 \, \text{rev/min})$$

Problem 5.4

A centrifuge with a phosphor-bronze basket 375 mm diameter is to be run at 30 Hz with a 100 mm layer of solids of bulk density 2000 kg/m³ at the walls. What should be the thickness of the walls of the basket if the perforations are so small that they have a negligible effect on strength?

Density of phosphor-bronze = 8900 kg/m³.

Maximum safe stress for phosphor-bronze = 55 MN/m².

Solution

The pressure exerted by the solids on the wall of the basket is given by:

$$P_c = 0.5\rho w^2 (b^2 - x^2)$$

where b is the basket radius (0·1875 m), x is the radius of inner surface of solids (0·1875 − 0·10) = 0·0875 m, w is the angular velocity (30 × 2π) = 60π rad/s, and ρ is the density of solids (2000 kg/m³).

$$\therefore \qquad P_c = 0.5 \times 2000(60\pi)^2 (0.1875^2 - 0.0875^2)$$

$$= 3.55 \times 10^7 (0.275)(0.10)$$

$$= 9.76 \times 10^5 \text{ N/m}^2$$

or $\qquad P_c = 0.98 \text{ MN/m}^2$

The stress in the walls is given by:

$$f = (b/\delta)(P_c + \rho_m \delta b w^2)$$

f will be taken as the maximum safe stress of phosphor-bronze, 55 × 10⁶ N/m²:

$$\rho_m = 8900 \text{ kg/m}^3$$

$$\therefore \qquad 55 \times 10^6 = (0.1875/\delta)[9.75 \times 10^5 + 8900\delta \times 0.1875(60\pi)^2]$$

$$\delta = 3.409 \times 10^{-9}(9.75 \times 10^5 + 5.929 \times 10^7 \delta)$$

$$= 3.323 \times 10^{-3} + 0.202\delta$$

$$\therefore \qquad \delta = 4.16 \times 10^{-3} \text{ m} \quad \text{or} \quad 4.16 \text{ mm}$$

In practice some safety margin would be allowed and a wall thickness of 5 mm would be specified.

Problem 5.5

An aqueous suspension consisting of particles of specific gravity 2·5 m in the size range 1–10 μm is introduced into a centrifuge with a basket 450 mm diameter rotating at 80 Hz. If the suspension forms a layer 75 mm thick in the basket, approximately how long will it take to cause the smallest particle to settle out?

Solution

Where the motion of the fluid with respect to the particle is turbulent, the time, taken for a particle to settle from h_1 to distance h_2 from the surface in a radial direction, is given by:

$$t = \frac{2}{a'}[(x+h_2)^{0\cdot5} - (x+h_1)^{0\cdot5}]$$

where $a' = \sqrt{[3dw^2(\rho_s - \rho)/\rho]}$, d is the diameter of the smallest particle $= 1 \times 10^{-6}$ m, w is the angular velocity of the basket $= (80 \times 2\pi) = 502\cdot7$ rad/s, ρ_s is the density of the solid $= 2500$ kg/m^3, ρ is the density of the fluid $= 1000$ kg/m^3, and x is the radius of the inner surface of the liquid $= 0\cdot150$ m.

∴ $a' = \sqrt{[3 \times 1 \times 10^{-6} \times 502\cdot7^2(2500 - 1000)/1000]}$

$$= 1\cdot066$$

and $t = (2/1\cdot066)[(0\cdot150 + 0\cdot075)^{0\cdot5} - (0\cdot150 + 0)^{0\cdot5}]$

$$= 1\cdot876(0\cdot474 - 0\cdot387)$$

$$= 0\cdot163\text{ s}$$

This is a very low value, equivalent to a velocity of $(0\cdot075/0\cdot163) = 0\cdot46$ m/s. Because of the very small diameter of the particle, it is more than likely that the conditions are streamline, even at this particle velocity.

For water, taking $\mu = 0\cdot001$ N s/m^2,

$$Re = 1 \times 10^{-6} \times 0\cdot46 \times 1000/0\cdot001 = 0\cdot46$$

and hence the following may be applied:

$t = \{18\mu/[d^2w^2(\rho_s - \rho)]\} \ln[(x+h_2)/(x+h_1)]$

$$= \{18 \times 0\cdot001/[10^{-12} \times 502\cdot7^2(2500 - 1000)]\} \ln[(0\cdot150 + 0\cdot075)/(0\cdot150 + 0)]$$

$$= 47\cdot5 \ln(0\cdot225/0\cdot150)$$

$$= \underline{\underline{19\cdot3\text{ s}}}$$

Problem 5.6

A centrifuge with a phosphor-bronze basket 375 mm diameter is to be run at 60 Hz with a 75 mm layer of liquid of specific gravity 1·2 in the basket. What thickness of walls is required in the basket?
Density of phosphor-bronze $= 8900$ kg/m^3.
Maximum safe working stress for phosphor-bronze $= 55$ MN/m^2.

Solution

The stress in the walls is given by:

$$f = (b/\delta)(P_c + \rho_m\, \delta b w^2)$$

where b is the radius of the basket (0·1875 m), δ is the thickness of the basket (m), ρ_m is the density of the basket material (8900 kg/m³), w is the angular velocity of the basket $= (60 \times 2\pi) = 377$ rad/s, P_c is the centrifugal pressure given by:

$$P_c = 0·5\rho w^2 (b^2 - x^2)$$

where ρ is the liquid density (1200 kg/m³), and x is the radius of the inner surface of fluid (0·1125 m).

$$\therefore \qquad P_c = 0·5 \times 1200 \times 377^2 (0·1875^2 - 0·1125^2)$$

$$= 8·527 \times 10^7 (0·3)(0·075)$$

$$= 1·92 \times 10^6 \text{ N/m}^2$$

Taking f as the maximum safe stress for phosphor-bronze, 55 MN/m² $= 55 \times 10^6$ N/m²:

$$55 \times 10^6 = (0·1875/\delta)(1·92 \times 10^6 + 8900\delta \times 0·1875 \times 377^2)$$

$$\therefore \qquad 2·93 \times 10^8 \delta = 1·92 \times 10^6 + 2·37 \times 10^8 \delta$$

$$\therefore \qquad \delta = 0·034 \text{ m} \quad \text{or} \quad 34 \text{ mm}$$

and the thickness to be specified allowing a reasonable margin would be

$$\underline{\underline{38·1 \text{ mm}}} \quad (1·5 \text{ in})$$

Problem 5.7

A centrifuge basket 600 mm long and 100 mm internal diameter has a discharge weir 25 mm diameter. What is the maximum volumetric flow of liquid through the centrifuge such that when the basket is rotated at 200 Hz all particles of diameter greater than 1 μm are retained on the centrifuge wall? The retarding force on a particle moving in a liquid can be taken as equal to $3\pi\mu du$, where u is the particle velocity relative to the liquid, μ is the liquid viscosity, and d is the particle diameter.
Sp. gr. of liquid $= 1·0$.
Sp. gr. of solid $= 2·0$.
Viscosity of liquid $(\mu) = 1·0$ mN s/m².
The inertia of the particle can be neglected.

Solution

With a basket radius of b m, the radius of the inner surface of liquid x m and h m the distance radially from the surface of the liquid, the equation of motion of a spherical particle of diameter d m under streamline conditions in the radial direction is:

$$(\pi d^3/6)(\rho_s - \rho)(x + h) w^2 - 3\pi\mu du - (\pi d^3/6) \rho_s \, du/dt = 0$$

Replacing u by dh/dt and neglecting the acceleration term:

$$(dh/dt) = d^2 (\rho_s - \rho) w^2 (x + h)/18\mu$$

The time any element of material remains in the basket is V'/Q, where Q is the volumetric rate of feed to the centrifuge and V' is the volume of liquid retained in the basket at any time. If the flow rate is so adjusted that a particle of diameter d is just retained when it has to travel through the maximum distance $h = (b - x)$ before reaching the wall,

$$h = d^2(\rho_s - \rho)bw^2V'/(18\mu Q)$$

or
$$Q = d^2(\rho_s - \rho)bw^2V'/(18\mu h)$$

In this case, $V' = (\pi/4)(0{\cdot}1^2 - 0{\cdot}025^2) \times 0{\cdot}6 = 0{\cdot}0044\,\text{m}^3$

$$h = (0{\cdot}10 - 0{\cdot}025/2)$$

$$\therefore\ Q = (1 \times 10^{-6})^2(2000 - 1000) \times 0{\cdot}1 \times (200 \times 2\pi)^2 \times 0{\cdot}0044/(18 \times 0{\cdot}001 \times 0{\cdot}0375)$$

$$\underline{\underline{= 1{\cdot}03 \times 10^{-3}\,\text{m}^3/\text{s}}}\quad (\text{approximately } 1\,\text{cm}^3/\text{s})$$

Problem 5.8

Calculate the minimum area and diameter of a thickener with a circular basin to treat $0{\cdot}1\,\text{m}^3/\text{s}$ of a slurry of solids concentration of $150\,\text{kg/m}^3$. The results of batch settling tests are as follows:

Solids concentration (kg/m³)	Settling velocity (μm/s)
100	148
200	91
300	55·33
400	33·25
500	21·40
600	14·50
700	10·29
800	7·38
900	5·56
1000	4·20
1100	3·27

A value of $1290\,\text{kg/m}^3$ for underflow concentration was selected from a retention time test. Estimate the underflow volumetric flow rate assuming total separation of all solids and that a clear overflow is obtained.

Solution

The settling velocity of the solids, $u\,\text{kg/m}^2\,\text{s}$, is calculated as

$$u = u_s c$$

where u_s is the settling velocity (m/s) and c the concentration of solids (kg/m³) and the data are plotted in Fig. 5a. From the point $u = 0$, $c = 1290\,\text{kg/m}^2$, a line is drawn which is just below the curve. This intercepts the axis at $u = 0{\cdot}0154\,\text{kg/m}^2\,\text{s}$.

The area of the thickener is then,

$$A = 0{\cdot}1 \times 150/0{\cdot}0154 = \underline{\underline{974\,\text{m}^2}}$$

and the diameter, $\qquad d = (4 \times 974/\pi)^{0 \cdot 5}$

$$= 35 \cdot 2 \, \text{m}$$

The volumetric flow rate of underflow is obtained from a mass balance as

$$= 0 \cdot 1 \times 150/1290$$

$$= \underline{\underline{0 \cdot 0116 \, \text{m}^3/\text{s}}}$$

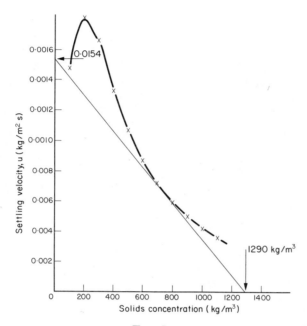

FIG. 5a

SECTION 6

FLUIDISATION

Problem 6.1

Oil of specific gravity 0·9 and viscosity 3 mN s/m² passes vertically upwards through a bed of catalyst consisting of approximately spherical particles of diameter 0·1 mm and specific gravity 2·6. At approximately what mass rate of flow per unit area of bed will (a) fluidisation, and (b) transport of particles occur?

Solution

(a) Use may be made of equations 4.9 and 6.1 to find the fluidising velocity, u_f.

$$u = \frac{1}{K''} \frac{e^3}{S^2(1-e)^2} \frac{1}{\mu} \frac{(-\Delta P)}{L} \qquad \text{(equation 4.9)}$$

$$-\Delta P = (1-e)(\rho_s - \rho)Lg \qquad \text{(equation 6.1)}$$

S = surface area/volume = $\pi d^2/(\pi d^3/6) = 6/d$ for a sphere.
Substituting $K'' = 5$, $S = 6/d$, and $-\Delta P/L$ from equation 6.1 into equation 4.9 gives:

$$u_f = 0.0055 \frac{e^3}{(1-e)} \frac{d^2(\rho_s - \rho)g}{\mu}$$

Hence

$$G_f' = \rho u = \frac{0.0055 e^3}{(1-e)} \frac{d^2 \rho(\rho_s - \rho)g}{\mu}$$

In this problem:

$$\rho_s = 2.6 \times 1000 = 2600 \, \text{kg/m}^3$$

$$\rho = 0.9 \times 1000 = 900 \, \text{kg/m}^3$$

$$\mu = 3.0 \times 10^{-3} \, \text{N s/m}^2$$

$$d = 0.1 \, \text{mm} = 1 \times 10^{-4} \, \text{m}$$

As no value for the voidage has been given, e will be calculated by considering eight close packed spheres of diameter d in a cube of side $2d$. Then:

$$\text{volume of spheres} = 8(\pi/6)d^3$$

$$\text{volume of enclosure} = (2d)^3 = 8d^3$$

∴   voidage e = $[8d^3 - 8(\pi/6)d^3]/8d^3 = 1 - (\pi/6) = 0.478$, say $e = 0.48$

Then $G_f' = 0.0055(0.48)^3(10^{-4})^2 \times 900 \times 1700 \times 9.81/(1-0.48) \times 3 \times 10^{-3}$

$$= 0.059 \, \text{kg/m}^2 \, \text{s}$$

(b) Transport of particles will occur when the fluid velocity is equal to the terminal falling velocity of the particle.

Using Stokes' law:

$$u_0 = d^2 g(\rho_s - \rho)/18\mu$$

$$= (10^{-4})^2 \times 9{\cdot}81 \times 1700/18 \times 3 \times 10^{-3}$$

$$\doteqdot 0{\cdot}0031 \text{ m/s}$$

Check Reynolds number $= 10^{-4} \times 0{\cdot}0031 \times 900/3 \times 10^{-3} = 0{\cdot}093$ (Stokes' law applies).

Hence required mass flow $= 0{\cdot}0031 \times 900$

$$= 2{\cdot}78 \text{ kg/m}^2\text{ s}$$

Alternatively, use may be made of Fig. 3.6 and equation 3.28:

$$(R/\rho u^2)\,Re^2 = 2d^3\rho g(\rho_s - \rho)/3\mu^2 \qquad \text{(equation 3.28)}$$

$$= 2 \times (10^{-4})^3 \times 900 \times 9{\cdot}81 \times 1700/3(3 \times 10^{-3})^2$$

$$= 1{\cdot}11$$

From Fig. 3.6, $\qquad\qquad Re = 0{\cdot}09$

Hence $\qquad u_0 = Re\mu/\rho d = 0{\cdot}09 \times 3 \times 10^{-3}/900 \times 10^{-4}$

$$= 0{\cdot}003 \text{ m/s}$$

and $\qquad\qquad G' = 0{\cdot}003 \times 900$

$$= 2{\cdot}7 \text{ kg/m}^2\text{ s}$$

Problem 6.2

Calculate the minimum velocity at which spherical particles (specific gravity 1·6) of diameter 1·5 mm will be fluidised by water in a tube of diameter 10 mm. Discuss the uncertainties in this calculation.

Viscosity of water $= 1\,\text{mN s/m}^2$; Kozeny's constant $= 5$.

Solution

As in Problem 6.1, the equation for the minimum fluidising velocity may be derived as:

$$u_f = 0{\cdot}0055\,\frac{e^3}{(1-e)}\,\frac{d^2(\rho_s - \rho)g}{\mu}$$

As a wall effect applies in this problem, use is made of equation 4.21 to find the correction factor f_w.

$$f_w = \left(1 + 0{\cdot}5\frac{Sc}{S}\right)^2$$

Sc = surface of the container/volume of bed

$$= (\pi \times 0{\cdot}01 \times 1)/(\pi/4)(0{\cdot}01)^2 \times 1 = 400\,\text{m}^2/\text{m}^3$$

$$S = 6/d \text{ for a spherical particle}$$

$$= 6/1 \cdot 5 \times 10^{-3} = 4000 \, \text{m}^2/\text{m}^3$$

Hence $$f_w = 1 \cdot 05$$

The uncertainty in this problem lies in the chosen value of the voidage e. If e is taken as 0·48 as in Problem 6.1:

$$u_f = 0 \cdot 0055 \times \frac{(0 \cdot 48)^3}{0 \cdot 52} \times (1 \cdot 5 \times 10^{-3})^2 \times \frac{(1600 - 1000)}{1 \times 10^{-3}} \times 9 \cdot 81$$

$$= 0 \cdot 0155 \, \text{m/s}$$

Allowing for wall effect, $u_f = 1 \cdot 05 \times 0 \cdot 0155 = \underline{0 \cdot 0163 \, \text{m/s}}$

If Ergun's equation is used to calculate the minimum fluidising velocity as illustrated fully in Problem 6.7, the value of u_f is found to be $\underline{0 \cdot 013 \, \text{m/s}}$.

Problem 6.3

In a fluidised bed, iso-octane vapour is adsorbed from an air stream on to the surface of alumina microspheres. The mol fraction of iso-octane in the inlet gas is $1 \cdot 442 \times 10^{-2}$ and the mol fraction in the outlet gas is found to vary with time in the following manner:

Time from start (s)	Mol fraction in outlet gas $(\times 10^2)$
250	0·223
500	0·601
750	0·857
1000	1·062
1250	1·207
1500	1·287
1750	1·338
2000	1·373

Show that the results can be interpreted on the assumptions that the solids are completely mixed, that the gas leaves in equilibrium with the solids, and that the adsorption isotherm is linear over the range considered. If the flow rate of gas is $0 \cdot 679 \times 10^{-6} \, \text{kmol/s}$ and the mass of solids in the bed is 4·66 g, calculate the slope of the adsorption isotherm. What evidence do the results provide concerning the flow pattern of the gas?

Solution

Work on mass transfer between fluid and particles is discussed in Chapter 6 of Vol. 2 where it is shown that by a mass balance over a bed of particles at any time t after the start of the experiment, the following equation is obtained:

$$G_m(y_0 - y) = \frac{d(WF)}{dt} \qquad \text{(equation 6.32)}$$

where G_m is the molar flow rate of gas, W is the mass of solids in the bed, F is the number of mols of vapour adsorbed on unit mass of solid, and y_0, y is the mol fraction of vapour in the inlet and outlet stream respectively.

If the adsorption isotherm is linear, and if equilibrium is reached between the outlet gas and the solids and if none of the gas bypasses the bed, then F is given by:

$$F = f + by$$

where f and b are the intercept and slope of the isotherm respectively.

Combining these equations and integrating gives:

$$\ln(1 - y/y_0) = -(G_m/Wb)t \qquad \text{(equation 6.36)}$$

If the assumptions outlined above are valid, a plot of $\ln(1 - y/y_0)$ against t should yield a straight line of slope $-G_m/Wb$. As $y_0 = 0.014\,42$, the following table may be produced.

Time (s)	y	y/y_0	$1 - y/y_0$	$\ln(1 - y/y_0)$
250	0.002 23	0.155	0.845	−0.168
500	0.006 01	0.417	0.583	−0.539
750	0.008 57	0.594	0.406	−0.902
1000	0.0106	0.736	0.263	−1.33
1250	0.0121	0.837	0.163	−1.81
1500	0.0129	0.893	0.107	−2.23
1750	0.0134	0.928	0.072	−2.63
2000	0.0137	0.952	0.048	−3.04

These data are plotted in Fig. 6a where a straight line is obtained. The slope is measured as $-0.001\,67/s$.

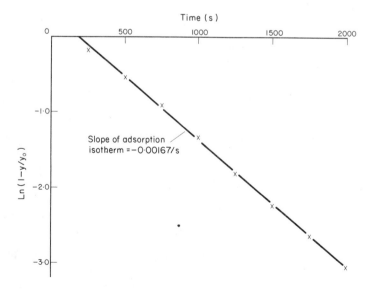

FIG. 6a

If $G_m = 0.679 \times 10^{-6}$ kmol/s and $W = 4.66$ g,

$$-0.001\ 67 = -0.679 \times 10^{-6}/4.66b$$

from which $b = 87.3 \times 10^{-6}$ kmol/g $\equiv 0.0873$ kmol/kg

Problem 6.4

Discuss the reasons for the good heat transfer properties of fluidised beds. Cold particles of glass ballotini are fluidised with heated air in a bed in which a constant flow of particles is maintained in a horizontal direction. When steady conditions have been reached, the temperatures recorded by a bare thermocouple immersed in the bed are as follows:

Distance above bed support (mm)	Temperature (K)
0	339·5
0·64	337·7
1·27	335·0
1·91	333·6
2·54	333·3
3·81	333·2

Calculate the coefficient for heat transfer between the gas and the particles, and the corresponding values of the particle Reynolds and Nusselt numbers. Comment on the results and on any assumptions made.

Gas flow rate = 0.2 kg/m² s.

Specific heat of air = 0.88 kJ/kg K.

Viscosity of air = 0.015 mN s/m².

Particle diameter = 0.25 mm.

Thermal conductivity of air = 0.03 W/m K.

Solution

Aspects of heat transfer are discussed in Chapter 6 of Vol. 2. For the system described in this problem, equation 6.43 relates the rate of heat transfer between particles and fluid by:

$$dQ = ha'\Delta t\ dz \qquad\qquad \text{(equation 6.43)}$$

and on integration

$$Q = ha' \int_0^z \Delta t\ dz \qquad\qquad \text{(equation 6.44)}$$

where Q is the heat transferred, h is the heat transfer coefficient, a' is the area for transfer/unit height of bed, and Δt is the temperature difference at height z.

From the data given, Δt may be plotted against z as shown in Fig. 6b where the area under the curve gives the value of the integral as 8·82 mm K.

Heat transferred $= 0.2 \times 0.88(339.5 - 332.2) = 1.11 \text{ kW/m}^2$ of bed cross-section
$$= 1100 \text{ W/m}^2$$
If bed voidage $= 0.57$, consider a bed $1 \text{ m}^2 \times 1 \text{ m}$ high, i.e. 1 m^3.
Volume of particles $= (1 - 0.57) \times 1 = 0.43 \text{ m}^3$.
Volume of 1 particle $= (\pi/6)(0.25 \times 10^{-3})^3 = 8.18 \times 10^{-12} \text{ m}^3$.
Therefore number of particles $= 0.43/8.18 \times 10^{-12} = 5.26 \times 10^{10}/\text{m}^3$.
Area of particles $= 5.26 \times 10^{10} \times (\pi/4)(0.25 \times 10^{-3})^2 = 1.032 \times 10^4 \text{ m}^2/\text{m}^3 = a'$.
Substituting in equation 6.44 gives:

$$1100 = h \times 1.03 \times 10^4 \times 8.82 \times 10^{-3}$$

$$h = \underline{\underline{12.2 \text{ W/m}^2 \text{ K}}}$$

From equation 6.46, $\quad Nu = 0.11 Re^{1.28}$

$$Re = G'd/\mu = 0.2 \times 0.25 \times 10^{-3}/0.015 \times 10^{-3}$$

$$= \underline{\underline{3.33}}$$

$$Nu = 0.11 \times (3.33)^{1.28} = \underline{\underline{0.513}}$$

$$h = 0.513 \times 0.03/0.25 \times 10^{-3}$$

$$= \underline{\underline{61.6 \text{ W/m}^2 \text{ K}}}$$

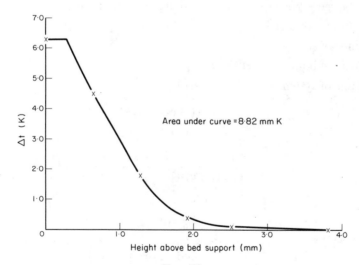

FIG. 6b

Problem 6.5

The relation between bed voidage e and fluid velocity u_c for particulate fluidisation of uniform particles, small compared with the diameter of the containing vessel, is given by:

$$\frac{u_c}{u_0} = e^n$$

where u_0 is the free-falling velocity.

Discuss the variation of the index n with flow conditions, indicating why it is independent of the Reynolds number Re with respect to the particle at very low and very high values of Re. When are appreciable deviations from this relation observed with liquid fluidised systems?

For particles of glass ballotini with free-falling velocities of 10 and 20 mm/s the index n has a value of 2·4. If a mixture of equal volumes of the two particles is fluidised, what will be the relation between the voidage and fluid velocity if it is assumed that complete segregation is obtained?

Solution

The variation of the index n with flow conditions is fully discussed in Chapters 5 and 6 of Vol. 2 under sedimentation and thickening and fluidisation. The ratio u_c/u_0 is in general dependent on the Reynolds number, voidage, and the ratio of particle diameter to that of the containing vessel. At low velocities, i.e. when $Re < 0·2$, the drag force is attributable entirely to skin friction, and at high velocities when $Re > 500$ skin friction becomes negligible and in these regions the ratio u_c/u_0 is independent of Re. For intermediate regions, equations 5.35 to 5.37 apply.

Consider unit volume of each particle, then:

Voidage of large particles $= e_1$, volume of liquid $= e_1/(1 - e_1)$.

Voidage of small particles $= e_2$, volume of liquid $= e_2/(1 - e_2)$.

Total volume of solids $= 2$.

Total volume of liquid $= e_1/(1 - e_1) + e_2/(1 - e_2)$.

Total volume of system $= 2 + e_1/(1 - e_1) + e_2/(1 - e_2)$.

$$\therefore \quad \text{voidage} = \frac{e_1/(1 - e_1) + e_2/(1 - e_2)}{2 + e_1/(1 - e_1) + e_2/(1 - e_2)}$$

$$= \frac{e_1(1 - e_2) + e_2(1 - e_1)}{2(1 - e_1)(1 - e_2) + e_1(1 - e_2) + e_2(1 - e_1)}$$

i.e.
$$e = \frac{e_1 + e_2 - 2e_1 e_2}{2 - e_1 - e_2}$$

Now
$$e_1 = \left(\frac{u}{u_{01}}\right)^{1/2·4} \quad \text{and} \quad e_2 = \left(\frac{u}{u_{01}/2}\right)^{1/2·4}$$

(since the free-falling velocities are in the ratio $1:2$)

$$\therefore \quad e_2 = e_1 2^{1/2·4}$$

$$\therefore \quad e = \frac{e_1 + e_1 \times 2^{1/2·4} - 2^{3·4/2·4} \times e_1^2}{2 - e_1 - 2^{1/2·4} \times e_1}$$

$$e = \frac{(u/20)^{1/2·4}(1 + 2^{1/2·4}) - 2^{3·4/2·4}(u/20)^{1/1·2}}{2 - (1 + 2^{1/2·4})(u/20)^{1/2·4}}$$

$$e = \frac{3u^{0.42} - u^{0.83}}{9 - 3u^{0.42}} \quad \text{(with } u \text{ in mm/s)}$$

$$\therefore \qquad 9e = 3eu^{0.42} = 3u^{0.42} - u^{0.84}$$

or

$$u^{0.84} - 3(1+e)u^{0.42} + 9e = 0$$

and

$$\underline{\underline{u^{0.42} = 1.5(1+e) + [2.24(1+e)^2 - 9e]}}$$

This relationship is plotted in Fig. 6c.

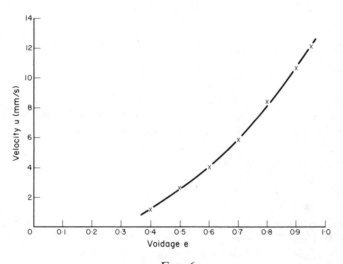

FIG. 6c

Problem 6.6

Obtain a relationship for the ratio of the terminal falling velocity of a particle to the minimum fluidising velocity for a bed of similar particles. Assume that Stokes' law and the Carman–Kozeny equation are applicable. What is the value of the ratio if the bed voidage at the minimum fluidising velocity is 0·4?

Discuss the validity of using the Carman–Kozeny equation for calculation of pressure drop through a fluidised bed.

Solution

Stokes' law (equation 3.17) and the Carman–Kozeny equation (equation 4.9) state respectively:

$$u_0 = d^2 g(\rho_s - \rho)/18\mu$$

and

$$u = \frac{1}{K''} \frac{e^3}{s^2(1-e)^2} \frac{1}{\mu} \frac{(-\Delta P)}{l}$$

In a fluidised bed the total frictional force must be equal to the effective weight of the bed. Then:

$$-\Delta P = (1-e)(\rho_s - \rho)lg \qquad \text{(equation 6.1)}$$

Substituting equation 6.1 into 4.9 and putting $K'' = 5$ gives:

$$u_f = 0.0055 \frac{e^3}{1-e} \frac{d^2(\rho_s - \rho)g}{\mu}$$

Hence
$$\frac{u_0}{u_f} = \frac{d^2 g(\rho_s - \rho)}{18\mu \times 0.0055} \frac{(1-e)}{e^3} \frac{\mu}{d^2(\rho_s - \rho)g}$$

$$= \frac{(1-e)}{18 \times 0.0055 e^3} = 10.1(1-e)/e^3$$

If $e = 0.4$, $u_0/u_f = \underline{\underline{94.7}}$

The use of the Carman–Kozeny equation is discussed in Section 4.2.3 of Chapter 4 of Vol. 2. (It is interesting to note that if $e = 0.48$, which was the value taken in Problem 6.1, $u_0/u_f = 47.5$, which agrees with the solution to that problem.)

Problem 6.7

A packed bed consisting of uniform spherical particles (diameter $d = 3\,\text{mm}$, density $\rho_s = 4200\,\text{kg/m}^3$) is fluidised by means of a liquid (viscosity $\mu = 1\,\text{mN s/m}^2$, density $\rho = 1100\,\text{kg/m}^3$). Using Ergun's equation for the pressure drop $(-\Delta P)$ through a bed height H and voidage e as a function of superficial velocity, calculate the minimum fluidising velocity in terms of the settling velocity (u_0) of the particles in the bed.

State clearly any assumptions which you make and indicate how closely you would expect your results to be confirmed by an experiment.

Ergun's equation:

$$-\frac{\Delta P}{H} = 150 \frac{(1-e)^2}{e^3} \frac{\mu u}{d^2} + 1.75 \frac{(1-e)}{e^3} \frac{\rho u^2}{d}$$

Solution

Equation 6.1 gives the pressure drop through a fluidised bed of height H as:

$$-\Delta P/H = (1-e)(\rho_s - \rho)g \qquad \text{(equation 6.1)}$$

Writing $u = u_f =$ fluidising velocity in Ergun's equation and substituting for $-\Delta P/H$ gives:

$$(1-e)(\rho_s - \rho)g = \frac{150(1-e)^2/\mu u_f}{e^3 d^2} + \frac{1.75(1-e)\rho u_f^2}{e^3 d}$$

or
$$(\rho_s - \rho)g = \frac{150(1-e)\mu u_f}{e^3 d^2} + \frac{1.75\rho u_f^2}{e^3 d}$$

If $d = 3 \times 10^{-3}\,\text{m}$, $\rho_s = 4200\,\text{kg/m}^3$, $\rho = 1100\,\text{kg/m}^3$, $\mu = 1 \times 10^{-3}\,\text{N s/m}^2$, and if e

is taken as 0·48 as in Problem 6.1, these figures may be substituted into the above equation to give:

$$3·04 = 7·84u_f + 580u_f^2$$

or $$u_f = \underline{0·066\,\text{m/s}} \quad \text{neglecting the negative root.}$$

If Stokes' law applies,

$$u_0 = d^2g(\rho_s - \rho)/18\mu$$
$$= (3 \times 10^{-3}) \times 9·81 \times 3100/18 \times 10^{-3}$$
$$= 15·21\,\text{m/s}$$

Then the value of Re

$$= 3 \times 10^{-3} \times 15·21 \times 4200/10^{-3}$$
$$= 1·92 \times 10^5$$

which is outside the range of Stokes' law. A Reynolds number of this order corresponds to region (c) of Fig. 3.5 where:

$$u_0^2 = 3dg(\rho_s - \rho)/\rho$$
$$= 3 \times 3 \times 10^{-3} \times 9·81 \times 3100/1100$$

and $$u_0 = 0·5\,\text{m/s}$$

A check on Re gives

$$Re = 3 \times 10^{-3} \times 0·5 \times 4200/10^{-3}$$
$$= 6·3 \times 10^3$$

which is within the limits of region (c).

Hence $$u_0/u_f = 0·5/0·066 = \underline{7·5}$$

As mentioned in Problem 6.2, empirical relationships for the minimum fluidising velocity are presented as a function of Reynolds number and this problem illustrates the importance of using the equations applicable to the particle Reynolds number in question.

PNEUMATIC AND HYDRAULIC CONVEYING

Problem 7.1

It is required to transport sand of particle size 1·25 mm and density 2600 kg/m³ at the rate of 1 kg/s through a horizontal pipe 200 m long. Estimate the air flow rate required, the pipe diameter, and the pressure drop in the pipeline.

Solution

For conventional pneumatic transport in pipelines, a solids–gas mass ratio of about 5 is employed.

$$\therefore \qquad \text{mass flow of air} = 0\cdot20\,\text{kg/s}$$

and, taking the density of air as 1·0 kg/m³,

$$\text{volumetric flow rate of air} = \underline{\underline{0\cdot20\,\text{m}^3/\text{s}}}\ .$$

In order to avoid excessive pressure drops, an air velocity of 30 m/s seems not un-reasonable.

Ignoring the volume occupied by the sand (which is about 0·2% of that occupied by the air), the cross-sectional area of pipe required

$$= 0\cdot20/30 = 0\cdot0067\,\text{m}^2$$

equivalent to a pipe diameter of $\sqrt{(4 \times 0\cdot0067/\pi)} = 0\cdot092$ m or 92 mm.

Thus a pipe diameter of $\underline{101\cdot6\ \text{mm}}$ (4 in) would be specified.

For sand of particle size $\overline{1\cdot25\,\text{mm}}$ and density 2600 kg/m³, the free-falling velocity u_0 is 4·7 m/s (Table 7.1).

In equation 7.1,

$$u_a - u_s = 4\cdot7/[0\cdot468 + 7\cdot25\,\sqrt{(4\cdot7/2600)}] = 6\cdot05\,\text{m/s}$$

The cross-sectional area of a 101·6 mm i.d. pipe $= \pi \times 0\cdot1016^2/4 = 0\cdot0081\,\text{m}^2$.

$$\therefore \qquad \text{air velocity } u_a = 0\cdot20/0\cdot0081 = 24\cdot7\,\text{m/s}$$

$$\therefore \qquad u_s = 24\cdot7 - 6\cdot05 = 18\cdot65\,\text{m/s}$$

Taking the viscosity and density of air as $1\cdot7 \times 10^{-5}$ N s/m² and 1·0 kg/m³ respectively, the Reynolds number for the air flow alone is:

$$Re = 0\cdot102 \times 24\cdot7 \times 1\cdot0/1\cdot7 \times 10^{-5} = 1\cdot48 \times 10^5$$

and from Fig. 3.8 (Vol. 1), the friction factor $f = 0\cdot004$.

In equation 3.20 (Vol. 1)

$$-\Delta P_{air} = 4f(l/d)\rho u^2/2$$

$$= 4 \times 0.004(200/0.102) \times 1.0 \times 24.7^2/2$$

$$= 9570 \text{ N/m}^2 \quad \text{or} \quad 9.57 \text{ kN/m}^2$$

assuming isothermal conditions and incompressible flow.

In equation 7.2,

$$(-\Delta P_x/-\Delta P_{air})(u_s^2/F) = 2805/u_0$$

$$\therefore \qquad -\Delta P_x = (2805\Delta P_{air}F)/(u_0 u_s^2)$$

$$= 2805 \times 9.57 \times 1.0/(4.7 \times 18.65^2)$$

$$= \underline{\underline{16.4 \text{ kN/m}^2}}$$

Problem 7.2

Sand of a mean diameter 0.2 mm is to be conveyed in water flowing at 0.5 kg/s in a 25 mm i.d. horizontal pipe 100 m long. What is the maximum amount of sand which may be transported in this way if the head developed by the pump is limited to 300 kN/m²? Assume fully suspended heterogeneous flow.

Solution

The settling velocity may be obtained from equation 3.28, assuming a spherical shape. Taking the density of sand as 2600 kg/m³ and the density and viscosity of water as 1000 kg/m³ and 0.001 N s/m² respectively:

$$(R_0'/\rho u_0^2) Re_0'^2 = [2 \times 0.0002^3 \times 1000(2600 - 1000)9.81]/(3 \times 0.001^2)$$

$$= 83.7$$

$$\therefore \qquad \log_{10} 83.7 = 1.923$$

and $\qquad \log_{10} Re_0' = 0.679 \quad$ (from Table 3.2)

$$Re_0' = 4.773$$

$$\therefore \qquad u_0 = 4.773 \times 0.001/(0.0002 \times 1000)$$

$$= 0.0239 \text{ m/s}$$

The mean velocity of the suspension (taken as that of the water)

$$u_m = 0.5/(1000\pi \, 0.025^2/4)$$

$$= 1.02 \text{ m/s}$$

For water alone, flowing at 1.02 m/s,

$$Re = 0.025 \times 1.02 \times 1000/0.001$$

$$= 25,500$$

From Fig. 3.8 (Vol. 1),

$$f = 0\cdot0092 \quad \text{(assuming } e/d = 0\cdot008)$$

In equation 3.21 (Vol. 1),

$$h_f = 4 \times 0\cdot0092(100/0\cdot025)(1\cdot02^2/2 \times 9\cdot81)$$

$$= 7\cdot8 \text{ m water}$$

$$\therefore \qquad i_w = 7\cdot8/100 = 0\cdot078 \text{ m water/m}$$

A pressure drop of $300\,\text{kN/m}^2$ or $3 \times 10^5\,\text{N/m}^2$ is equivalent to

$$3 \times 10^5/(1000 \times 9\cdot81) = 30\cdot6 \text{ m water}$$

$$\therefore \qquad i = 30\cdot6/100 = 0\cdot306 \text{ m water/m}$$

In equation 7.8:

$$(0\cdot306 - 0\cdot078)/0\cdot078C = 1100(9\cdot81 \times 0\cdot025/1\cdot02^2)(0\cdot0239/1\cdot02)(2\cdot6 - 1)$$

$$\therefore \qquad C = 0\cdot30$$

It is of interest to compare this result with that obtained from equation 7.10. In this case,

$$(0\cdot306 - 0\cdot078)/0\cdot078C$$

$$= 121\langle[9\cdot81 \times 0\cdot025(2\cdot6 - 1)0\cdot0239]/\{1\cdot02^2[9\cdot81 \times 0\cdot0002(2\cdot6 - 1)]^{0\cdot5}\}\rangle^{1\cdot5}$$

and $C = 0\cdot36$, which is a very similar result.]
  If q kg/s is the mass flow of sand,

$$\text{volumetric flow of sand} = q/2600 = 0\cdot000\,385 \text{ m}^3/\text{s}$$

$$\text{volumetric flow of water} = 0\cdot5/1000 = 0\cdot0005 \text{ m}^3/\text{s}$$

$$\therefore \qquad 0\cdot000\,385q/(0\cdot000\,385q + 0\cdot0005) = 0\cdot30$$

and $$\qquad \underline{\underline{q = 0\cdot56\,\text{kg/s}}}$$

GAS CLEANING

Problem 8.1

The size distribution by weight of the dust carried in a gas is given in the following table, together with the efficiency of collection over each size range.

Size range (μm)	0–5	5–10	10–20	20–40	40–80	80–160
Per cent weight	10	15	35	20	10	10
Per cent efficiency	20	40	80	90	95	100

Calculate the overall efficiency of the collector and the percentage by weight of the emitted dust which is smaller than 20 μm in diameter.

If the dust burden is 18 g/m^3 at entry and the gas flow 0·3 m^3/s, calculate the weight of dust emitted in kg/s.

Solution

Taking 1 m^3 of gas as the basis of calculation, the table given in the statement of the problem may be completed as the inlet dust concentration is 18 g/m^3. Thus:

Size range (μm)	0–5	5–10	10–20	20–40	40–80	80–160	Total
Weight in each size range (g)	1·8	2·7	6·3	3·6	1·8	1·8	18·0
Weight retained (g)	0·36	1·08	5·04	3·24	1·71	1·80	13·23
Weight emitted (g)	1·44	1·62	1·26	0·36	0·09	0	4·77

Hence, overall efficiency $= (13\cdot23/18) \times 100 = \underline{73\cdot5\%}$.

% emitted $< 20\,\mu$m $= [(1\cdot44 + 1\cdot62 + 1\cdot26)/4\cdot77] \times 100 = \underline{90\cdot1\%}$.

Inlet gas flow $= 0\cdot3\,$m^3/s.

$\therefore$ weight emitted $= 0\cdot3 \times 4\cdot77 = 1\cdot43\,$g/s

$$= \underline{1\cdot43 \times 10^{-3}\,\text{kg/s}} \quad (0\cdot12\,\text{tonne/day})$$

Problem 8.2

The collection efficiency of a cyclone is 45% over the size range 0–5 μm, 80% over the size range 5–10 μm, and 96% for particles exceeding 10 μm. Calculate the efficiency of collection for the following dust:

Weight distribution: 50% 0–5 μm

                      30% 5–10 μm

                      20% above 10 μm

Solution

For the collector: Size (μm)	0–5	5–10	>10	
Efficiency (%)	45	80	96	
For the dust: Weight (%)	50	30	20	
Basis 100 kg dust, weight at inlet (kg)	50	30	20	
Weight retained (kg)	22·5	24·0	19·2	Total = 65·7 kg

Overall efficiency $= (65\cdot7/100) \times 100 = \underline{\underline{65\cdot7\%}}$

Problem 8.3

A sample of dust from the air in a factory has been collected on a glass slide. If the dust on the slide was deposited from one cubic centimetre of air, estimate the weight of dust in grams per cubic metre of air in the factory, given the number of particles in the various size ranges to be shown in the following table:

Size range (μm)	0–1	1–2	2–4	4–6	6–10	10–14
Number of particles	2000	1000	500	200	100	40

Assume the specific gravity of the dust to be 2·6, and make an appropriate allowance for particle shape.

Solution

It will be assumed that the particles are spherical so that if the particle diameter is d m and the density $\rho = 2600$ kg/m^3,

$$\text{volume of 1 particle} = (\pi/6)\, d^3 \text{ m}^3$$
$$\text{weight of 1 particle} = 2600\,(\pi/6)\, d^3 \text{ kg}$$

The following table may then be produced.

Size (μm)	0–1	1–2	2–4	4–6	6–10	10–14
Number of particles	2000	1000	500	200	100	40
Average dia. (μm)	0·5	1·5	3·0	5·0	8·0	12·0
Diameter (m)	$0\cdot5 \times 10^{-6}$	$1\cdot5 \times 10^{-6}$	$3\cdot0 \times 10^{-6}$	$5\cdot0 \times 10^{-6}$	$8\cdot0 \times 10^{-6}$	$12\cdot0 \times 10^{-6}$
Volume (m^3)	$6\cdot54 \times 10^{-20}$	$3\cdot38 \times 10^{-18}$	$1\cdot41 \times 10^{-17}$	$6\cdot54 \times 10^{-17}$	$2\cdot68 \times 10^{-16}$	$9\cdot05 \times 10^{-16}$
Weight of 1 particle (kg)	$1\cdot70 \times 10^{-16}$	$8\cdot78 \times 10^{-15}$	$3\cdot68 \times 10^{-14}$	$1\cdot70 \times 10^{-13}$	$6\cdot97 \times 10^{-13}$	$2\cdot35 \times 10^{-12}$
Weight of particles in size range (kg)	$3\cdot40 \times 10^{-13}$	$8\cdot78 \times 10^{-12}$	$1\cdot83 \times 10^{-11}$	$3\cdot40 \times 10^{-11}$	$6\cdot97 \times 10^{-11}$	$9\cdot41 \times 10^{-11}$

Total weight of particles $= 2{\cdot}50 \times 10^{-10}\,\text{kg}$.

As this weight is obtained from $1\,\text{cm}^3$ of air, required dust concentration is given by:

$$2{\cdot}50 \times 10^{-10} \times 10^3 \times 10^6 = \underline{\underline{0{\cdot}25\,\text{g/m}^3}}$$

Problem 8.4

A cyclone separator, $0{\cdot}3\,\text{m}$ in diameter and $1{\cdot}2\,\text{m}$ long, has a circular inlet $25\,\text{mm}$ in diameter and an outlet of the same size. If the gas enters at $1{\cdot}5\,\text{m/s}$, at what particle size will the theoretical cut occur?

Viscosity of air $= 0{\cdot}018\,\text{mN s/m}^2$.

Density of air $= 1{\cdot}3\,\text{kg/m}^3$.

Density of particles $= 2700\,\text{kg/m}^3$.

Solution

The radius at which a particle will rotate within the body of a cyclone corresponds to the position where the net radial force on the particle is zero. The two forces acting are the centrifugal force outwards and the frictional drag of the gas acting inwards.

Consider a spherical particle of diameter d rotating at a radius r. Then the centrifugal force is:

$$\frac{mu_t^2}{r} = \frac{(\pi/6)\,d^3\rho_s\,u_t^2}{r} \qquad \text{(equation 8.1)}$$

where m is the mass of the particle and u_t is the tangential component of the velocity of the gas.

It is assumed here that there is no slip between the gas and the particle in the tangential direction.

If the radial velocity is low, the inward radial force due to friction will from equation 3.1 be equal to $3\pi\mu du_r$, where μ is the viscosity of the gas and u_r is the radial component of the velocity of the gas.

The radius r, at which the particle will rotate at equilibrium, is then given by

$$\frac{(\pi/6)\,d^3\rho_s\,u_t^2}{r} = 3\pi\mu du_r$$

i.e.

$$\frac{u_t^2}{r} = \frac{18\mu}{d^2\rho_s}u_r \qquad \text{(equation 8.2)}$$

Now the free-falling velocity of the particle u_0 is given by equation 3.28, and when the density of the particle is large compared with that of the gas,

$$u_0 = \frac{d^2 g\rho_s}{18\mu} \qquad \text{(equation 8.3)}$$

Substituting in equation 8.2:

$$\frac{u_t^2}{r} = \frac{u_r}{u_0} g$$

or

$$u_0 = \frac{u_r}{u_t^2} rg \qquad\qquad \text{(equation 8.4)}$$

Thus the higher the terminal falling velocity of the particle, the greater is the radius at which it will rotate and the easier it is to separate. If it is assumed that a particle will be separated provided it tends to rotate outside the central core of diameter $0.4d_0$, the terminal falling velocity of the smallest particle which will be retained is found by substituting $r = 0.2d_0$ in equation 8.4,

i.e.

$$u_0 = \frac{u_r}{u_t^2} 0.2d_0\, g \qquad\qquad \text{(equation 8.5)}$$

In order to calculate u_0 it is necessary to evaluate u_r and u_t for the region outside the central core. The radial velocity u_r is found to be approximately constant at a given radius and to be given by the volumetric rate of flow of gas divided by the cylindrical area for flow at the radius r. Thus, if G is the mass rate of flow of gas through the separator and ρ is its density, the linear velocity in a radial direction at a distance r from the centre is given by

$$u_r = \frac{G}{2\pi r Z \rho} \qquad\qquad \text{(equation 8.6)}$$

where Z is the depth of the separator.

The tangential velocity is found experimentally to be inversely proportional to the square root of the radius at all depths. Then if u_t is the tangential component of the velocity at a radius r, and u_{t0} is the corresponding value at the circumference of the separator:

$$u_t = u_{t0} \sqrt{\frac{d_t}{2r}} \qquad\qquad \text{(equation 8.7)}$$

Further, it is found that u_{t0} is approximately equal to the velocity with which the gas stream enters the cyclone separator. If these values for u_r and u_t are now substituted into equation 8.5, the terminal falling velocity of the smallest particle which the separator will retain is given by:

$$u_0 = \frac{G}{2\pi \times 0.2d_0\, \rho Z} \frac{2 \times 0.2d_0}{d_t} \frac{1}{u_{t0}^2} 0.2d_0\, g$$

$$= \frac{0.2Gd_0\, g}{\pi \rho Z d_t\, u_{t0}^2} \qquad\qquad \text{(equation 8.8)}$$

If the cross-sectional area of the inlet is A_i, $G = A_i \rho u_{t0}$

and

$$u_0 = \frac{0.2A_i^2 d_0\, \rho g}{\pi Z d_t\, G} \qquad\qquad \text{(equation 8.9)}$$

Using data provided in the statement of the problem:

A_i = cross-sectional area at gas inlet = $(\pi/4)(0{\cdot}075)^2$

$$= 4{\cdot}42 \times 10^{-3}\,\text{m}^2$$

d_o = gas outlet diameter = $0{\cdot}075\,\text{m}$

ρ = gas density = $1{\cdot}30\,\text{kg/m}^3$

Z = height of separator = $1{\cdot}2\,\text{m}$

d_t = separator diameter = $0{\cdot}3\,\text{m}$

G = mass flow rate of gas = $1{\cdot}5 \times 4{\cdot}42 \times 10^{-3} \times 1{\cdot}3$

$$= 8{\cdot}62 \times 10^{-3}\,\text{kg/s}$$

$$\therefore \quad u_0 = \frac{0{\cdot}2 \times (4{\cdot}42 \times 10^{-3})^2 \times 0{\cdot}075 \times 1{\cdot}3 \times 9{\cdot}81}{\pi \times 1{\cdot}2 \times 0{\cdot}3 \times 8{\cdot}62 \times 10^{-3}}$$

$$= 3{\cdot}83 \times 10^{-4}\,\text{m/s}$$

Use is now made of Stokes' law to find the particle diameter:

$$u_0 = d^2 g(\rho_s - \rho)/18\mu$$

$$\therefore \quad d = [u_0 \times 18\mu/g(\rho_s - \rho)]^{0{\cdot}5}$$

$$= [3{\cdot}83 \times 10^{-4} \times 18 \times 0{\cdot}018 \times 10^{-3}/9{\cdot}81(2700 - 1{\cdot}3)]^{0{\cdot}5}$$

$$= 2{\cdot}17 \times 10^{-6}\,\text{m}$$

$$= \underline{\underline{2{\cdot}17\,\mu\text{m}}}$$

FILTRATION

Problem 9.1

A slurry containing 0·2 kg solid (specific gravity 3·0) per kilogram of water is fed to a rotary drum filter 0·6 m long and 0·6 m diameter. The drum rotates at one revolution in 350 s and 20% of the filtering surface is in contact with the slurry at any instant. If filtrate is produced at the rate of 0·125 kg/s and the cake has a voidage of 0·5, what thickness of cake is produced when filtering at an absolute pressure of 35 kN/m²?

The rotary filter breaks down and the operation has to be carried out temporarily in a plate and frame press with frames 0·3 m square. The press takes 100 s to dismantle and 100 s to reassemble and, in addition, 100 s is required to remove the cake from each frame. If filtration is to be carried out at the same overall rate as before, with an operating pressure of 275 kN/m², what is the minimum number of frames that need to be used and what is the thickness of each? Assume the cakes to be incompressible and neglect the resistance of the filter media.

Solution

Drum filter

Area of filtering surface $= 0·6 \times 0·6\pi = 0·36\pi\,\text{m}^2$

Rate of filtration $= 0·125\,\text{kg/s}$

$$= 0·125/1000 = 1·25 \times 10^{-4}\,\text{m}^3/\text{s of filtrate}$$

Volumetric rate of deposition of solids (bulk)

$$= 1·25 \times 10^{-4} \times 0·2/(0·5 \times 3·0) = 1·67 \times 10^{-5}\,\text{m}^3/\text{s}$$

One revolution takes 350 s; therefore a given piece of filtering surface is immersed for $(350 \times 0·2) = 70\,\text{s}$.

Bulk volume of cake deposited per revolution

$$= 1·67 \times 10^{-5} \times 350 = 5·85 \times 10^{-3}\,\text{m}^3$$

Thickness of cake produced

$$= 5·85 \times 10^{-3}/0·36\pi = 5·17 \times 10^{-3}\,\text{m}$$

or $\underline{5·2\,\text{mm}}$

Now $\dfrac{dV}{dt} = \dfrac{(-\Delta P)A^2}{r\mu l} = \dfrac{(-\Delta P)A^2}{r\mu V_v}$ (from equation 9.2)

At constant pressure, $V^2 = (2/r\mu v)(-\Delta P)A^2t = K(-\Delta P)A^2t$ (say) (from equation 9.12)

Then, expressing pressures, areas, times, and volumes in kN/m^2, m^2, s, and m^3 respectively, for one revolution of the drum,

$$(1{\cdot}25 \times 10^{-4} \times 350)^2 = K(101{\cdot}3 - 35)(0{\cdot}36\pi)^2 \times 70$$

(since each element of area is immersed for one-fifth of cycle),

i.e.
$$K = 3{\cdot}22 \times 10^{-7}$$

Filter press

Use a filter press with n frames of thickness d m.

Total time for one complete cycle of press $= t_f + 100n + 200$ s, where t_f is the time during which filtration is occurring.

Overall rate of filtration $= V_f/(t_f + 100n + 200) = 1{\cdot}25 \times 10^{-4}\,m^3/s$, where V_f is the total volume of filtrate per cycle.

Now V_f = volume of frames/volume of cake deposited by unit volume of filtrate (v)

$$= 0{\cdot}3^2 nd/[0{\cdot}2/(0{\cdot}5 \times 3{\cdot}0)] = 0{\cdot}675nd$$

But
$$V_f = 3{\cdot}22 \times 10^{-7}(275 - 101{\cdot}3)(2n \times 0{\cdot}3 \times 0{\cdot}3)^2 \times t_f = (0{\cdot}675nd)^2$$

i.e.
$$t_f = 2{\cdot}516 \times 10^5 d^2$$

Thus
$$1{\cdot}25 \times 10^{-4} = 0{\cdot}675nd/(2{\cdot}516 \times 10^5 d^2 + 100n + 200)$$

i.e.
$$31{\cdot}45d^2 + 0{\cdot}0125n + 0{\cdot}0250 = 0{\cdot}675nd$$

giving
$$n = \frac{0{\cdot}0250 + 31{\cdot}45d^2}{0{\cdot}675d - 0{\cdot}0125}$$

n is a minimum when $dn/dd = 0$,

i.e. when
$$(0{\cdot}675d - 0{\cdot}0125) \times 62{\cdot}9d - (0{\cdot}0250 + 31{\cdot}45d^2) \times 0{\cdot}675 = 0$$

$$d^2 - 0{\cdot}0370d - 0{\cdot}000\,796 = 0$$

$$d = 0{\cdot}0185 \pm \sqrt{(0{\cdot}000\,343 + 0{\cdot}000\,796)}$$

$$= 0{\cdot}0522\,m \quad \text{or} \quad \underline{52{\cdot}2\,mm}$$

Hence
$$n = (0{\cdot}0250 + 31{\cdot}45 \times 0{\cdot}0522^2)/(0{\cdot}675 \times 0{\cdot}0522 - 0{\cdot}0125)$$

$$= 4{\cdot}87$$

Thus a minimum of 5 frames must be used.

The size of frames which will give exactly the required rate of filtration when five are used are given by,
$$0{\cdot}0250 + 31{\cdot}45d^2 = 3{\cdot}375d - 0{\cdot}0625$$

i.e.
$$d^2 - 0{\cdot}107d + 0{\cdot}002\,78 = 0$$

$$d = 0{\cdot}0535 \pm \sqrt{(0{\cdot}002\,85 - 0{\cdot}002\,78)}$$

$$= 0{\cdot}044\,m \quad \text{or} \quad 0{\cdot}063\,m$$

i.e. 5 frames of thickness 44 mm or 63 mm will give exactly the required filtration rate; intermediate sizes give higher rates.

Thus any frame thickness between 44 and 63 mm will be satisfactory. In practice 2 in (50·4 mm) frames would be used.

Problem 9.2

A slurry containing 100 kg of whiting, of specific gravity 3·0, per m³ of water is filtered in a plate and frame press, which takes 900 s to dismantle, clean, and reassemble. If the filter cake is incompressible and has a voidage of 0·4, what is the optimum thickness of cake for a filtration pressure of 1000 kN/m²? If the cake is washed at 500 kN/m² and the total volume of wash water employed is one-quarter of that of the filtrate, how is the optimum thickness of the cake affected? Neglect the resistance of the filter medium and take the viscosity of water as 1 mN s/m². In an experiment, a pressure of 165 kN/m² produced a flow of water of 0·06 cm³/s through a centimetre cube of filter cake.

Solution

The basic filtration equation may be written:

$$\frac{1}{A}\frac{dV}{dt} = \frac{(-\Delta P)}{r\mu l} \qquad \text{(equation 9.2)}$$

r is defined as the specific resistance of the cake, and using the data given may be calculated for the flow through a cube of cake.

$\Delta P = 165 - 101\cdot3 = 63\cdot7\,\text{kN/m}^2 = 63\cdot7 \times 10^3\,\text{N/m}^2.$

$A = 1\,\text{cm}^2.$

$l = 1\,\text{cm}.$

$\mu = 1 \times 10^{-3}\,\text{N s/m}^2.$

$dV/dt = 0\cdot02\,\text{cm}^3/\text{s}.$

Hence
$$r = \frac{63\cdot7 \times 10^3 \times 1}{1 \times 10^{-3} \times 1 \times 0\cdot02} = 3185 \times 10^6/\text{cm}^2$$

$$= 3185 \times 10^{10}/\text{m}^2$$

The slurry contains 100 kg whiting/m³ of water.

Volume of 100 kg whiting $= 100/3000 = 0\cdot0333\,\text{m}^3.$

Volume of cake $= 0\cdot0333/(1 - 0\cdot4) = 0\cdot0556\,\text{m}^3.$

Volume of liquid in cake $= 0\cdot0333 \times 0\cdot4/0\cdot6 = 0\cdot0222\,\text{m}^3.$

Volume of filtrate $= 1 - 0\cdot0222 = 0\cdot978\,\text{m}^3.$

∴ $v = $ volume of cake/volume of filtrate $= 0\cdot056$

From equation 9.2, $$V^2 = \frac{2A^2(-\Delta P)t}{r\mu v}$$

But $L = $ half frame thickness $= Vv/A$ (equation 9.6)

∴ $V^2 = L^2 A^2/v^2$

and
$$L^2 = \frac{2A(-\Delta P)vt}{r\mu}$$

$$= \frac{2 \times (1000 - 101 \cdot 3) \times 10^3 \times 0 \cdot 056t}{3185 \times 10^{10} \times 1 \times 10^{-3}}$$

$$L^2 = 3 \cdot 16 \times 10^{-6}t$$

It is shown in section 9.4.3 that if the resistance of the filter medium is neglected, the optimum cake thickness occurs when the filtration time is equal to the downtime,

i.e. $\qquad\qquad$ for $\quad L_{opt}, \quad t = 900$

$\therefore \qquad\qquad L^2_{opt} = 3 \cdot 16 \times 10^{-6} \times 900 = 2 \cdot 84 \times 10^{-3}$

and $\qquad\qquad\qquad L = 0 \cdot 053 \, \text{m}$

$\therefore \qquad\qquad$ optimum frame thickness $= \underline{\underline{106 \, \text{mm}}}$

For the washing process, if the filtration pressure is halved, the rate of washing is halved. The wash water has twice the thickness to penetrate and half the area for flow that is available to the filtrate, so that, considering these factors, the washing rate is one-eighth of the final filtration rate.

The final filtrate $\qquad\qquad \dfrac{dV}{dt} = \dfrac{A^2(-\Delta P)}{r\mu v V}$

$$= \frac{898 \cdot 7 \times 10^3 A^2}{3185 \times 10^{10} \times 10^{-3} \times 0 \cdot 056V} = \frac{5 \cdot 04 \times 10^{-4} A^2}{V}$$

$\therefore \qquad\qquad$ washing rate $=$ final rate/8 $= 63 \times 10^{-5} A^2/V$

The volume of wash water $= V/4$.

Hence $\qquad\qquad$ washing time $= (V/4)/(6 \cdot 3 \times 10^{-5} A^2/V)$

i.e. $\qquad\qquad\qquad t_w = 3 \cdot 97 \times 10^3 V^2/A^2$

Now $\qquad\qquad\qquad V^2 = L^2 A^2/v^2$

$\therefore \qquad\qquad t_w = \dfrac{L^2 A^2}{(0 \cdot 056)^2} \times \dfrac{3 \cdot 97 \times 10^3}{A^2} = 1 \cdot 27 \times 10^6 L^2$

The filtration time t_f was shown earlier to be:

$$t_f = L^2/3 \cdot 16 \times 10^{-6} = 3 \cdot 16 \times 10^5 L^2$$

$\therefore \qquad$ total cycle time $= L^2(1 \cdot 27 \times 10^6 + 3 \cdot 16 \times 10^5) + 900$

$$= 1 \cdot 58 \times 10^6 L^2 + 900$$

The rate of cake production

$$= \frac{L}{1 \cdot 58 \times 10^6 L^2 + 900} = R$$

For $dR/dL = 0$, $1.58 \times 10^6 L^2 + 900 - 3.16 \times 10^6 L^2 = 0$

and $L = 0.024\,\text{m}$

$\therefore$ frame thickness $= 0.048\,\text{m} = \underline{\underline{48\,\text{mm}}}$

Problem 9.3

A plate and frame press, filtering a slurry, gave a total of $8\,\text{m}^3$ of filtrate in $1800\,\text{s}$ and $11\,\text{m}^3$ in $3600\,\text{s}$, when filtration was stopped. Estimate the washing time in seconds if $3\,\text{m}^3$ of wash water are used. The resistance of the cloth can be neglected and a constant pressure is used throughout.

Solution

For constant pressure filtration with no cloth resistance,

$$t = \frac{r\mu v}{2A^2(-\Delta P)}V^2 \qquad \text{(equation 9.12)}$$

At $t_1 = 1800\,\text{s}$, $V_1 = 8\,\text{m}^3$, and when $t_2 = 3600\,\text{s}$, $V_2 = 11\,\text{m}^3$

$\therefore$ $3600 - 1800 = \dfrac{r\mu v}{2A^2(-\Delta P)}(11^2 - 8^2)$

$$\frac{r\mu v}{2A^2(-\Delta P)} = 316$$

As $\dfrac{dV}{dt} = \dfrac{A^2(-\Delta P)}{r\mu v V}$

$$\frac{dV}{dt} = \frac{1}{2 \times 31.6V} = \frac{0.0158}{V}$$

The final rate of filtration $= 0.0158/11 = 1.44 \times 10^{-3}\,\text{m}^3/\text{s}$.

For thorough washing in a plate and frame filter the wash water has twice the thickness of cake to penetrate and half the area for flow that is available to the filtrate. Thus the flow of wash water at the same pressure will be one-quarter of the filtration rate.

Hence   rate of washing $= 1.44 \times 10^{-3}/4 = 3.6 \times 10^{-4}\,\text{m}^3/\text{s}$

time of washing $= 3/(3.6 \times 10^{-4})$

$$= \underline{\underline{8400\,\text{s}}} \quad (2.3\,\text{h})$$

Problem 9.4

In the filtration of a certain sludge the initial period is effected at a constant rate with the feed pump at full capacity till the pressure reaches $400\,\text{kN/m}^2$. The pressure is then maintained at this value for the remainder of the filtration. The constant rate operation requires 900 s, and one-third of the total filtrate is obtained during this period.

Neglecting the resistance of the filter medium, determine (a) the total filtration time, and (b) the filtration cycle with the existing pump for the maximum daily capacity, if the time for removing the cake and reassembling the press is 1200 s. The cake is not washed.

Solution

For filtration carried out at a constant filtration rate for time t_1 in which time a volume V_1 is collected and followed by a constant pressure period such that the total filtration time is t and the total volume of filtrate is V:

$$V^2 - V_1^2 = \frac{2A^2(-\Delta P)}{r\mu v}(t - t_1) \qquad \text{(equation 9.13)}$$

Assuming no cloth resistance,
For the constant rate period:

$$t_1 = \frac{r\mu v}{A^2(-\Delta P)} V_1^2 \qquad \text{(equation 9.10)}$$

Using the data given:

$$t_1 = 900\,\text{s}, \quad \text{volume} = V_1$$

$$\therefore \quad \frac{r\mu v}{A^2(-\Delta P)} = \frac{900}{V_1^2}$$

(a) For the constant pressure period,

$$V - V_1 = 2V_1 \quad \text{and} \quad t - t_1 = t_p$$

$$\therefore \quad (2V_1)^2 = \frac{2V_1^2}{900}t_p$$

$$\therefore \quad t_p = 1800\,\text{s}$$

$$\therefore \quad \text{total filtration time} = 900 + 1800 = \underline{\underline{2700\,\text{s}}}$$

$$\text{total cycle time} = 2700 + 1200 = \underline{\underline{3900\,\text{s}}}$$

(b) For the constant rate period,

$$t_1 = \frac{r\mu v}{A^2(-\Delta P)} V_1^2 = \frac{V_1^2}{K}$$

For the constant pressure period,

$$t - t_1 = \frac{r\mu v}{2A^2(-\Delta P)}(V^2 - V_1^2) = \frac{V^2 - V_1^2}{2K}$$

Total filtration time $\qquad = t = \dfrac{1}{K}\left(V_1^2 + \dfrac{V^2 - V_1^2}{2}\right)$

$$= \dfrac{(V^2 + V_1^2)}{2K}$$

Rate of filtration $\qquad = \dfrac{V}{t + t_d}$ where $t_d = $ downtime

$$= \dfrac{2KV}{V^2 + V_1^2 + 2Kt_d}$$

$\dfrac{d(\text{rate})}{dV} = 0$ for a maximum gives:

$$V_1^2 - V^2 + 2Kt_d = 0$$

or $\qquad t_d = \dfrac{1}{2K}(V^2 - V_1^2) = t - t_1$

Now $\qquad t_d = 1200\,\text{s} = t - 900$

$\therefore \qquad t = 2100\,\text{s}$

and $\qquad$ cycle time $= 2100 + 1200 = \underline{\underline{3300\,\text{s}}}$

Problem 9.5

A rotary filter, operating at 0·03 Hz, filters 0·0075 m³/s. Operating under the same vacuum and neglecting the resistance of the filter cloth, at what speed must the filter be operated to give a filtration rate of 0·0160 m³/s?

Solution

For constant pressure filtration in a rotary filter:

$$V^2 = \dfrac{A^2(-\Delta P)}{r\mu v}\,t$$

i.e. $\qquad V^2 \propto t \propto 1/N$

where N is the speed rotation.
As $V \propto 1/\sqrt{N}$ and the rate of filtration is V/t, then:

$$\dfrac{V}{t} \propto \left(\dfrac{1}{\sqrt{N}} \times \dfrac{1}{t}\right) \propto \left(\dfrac{1}{\sqrt{N}} \times N\right) \propto \sqrt{N}$$

$$\therefore \quad \frac{(V/t)_1}{(V/t)_2} = \frac{\sqrt{N_1}}{\sqrt{N_2}}$$

$$\therefore \quad \frac{0{\cdot}0075}{0{\cdot}0150} = \frac{\sqrt{0{\cdot}03}}{\sqrt{N_2}}$$

$$N_2 = \underline{0{\cdot}12\,\text{Hz}} \quad (7{\cdot}2\,\text{rev/min})$$

Problem 9.6

A slurry is filtered in a plate and frame press containing 12 frames, each $0{\cdot}3$ m square and 25 mm thick. During the first 200 s, the filtration pressure is slowly raised to the final value of $500\,\text{kN/m}^2$, and during this period the rate of filtration is maintained constant. After the initial period, filtration is carried out at constant pressure and the cakes are completely formed after a further 900 s. The cakes are then washed at $375\,\text{kN/m}^2$ for 600 s, using "thorough washing". What is the volume of filtrate collected per cycle and how much wash water is used?

A sample of the slurry had previously been tested, using a vacuum leaf filter of $0{\cdot}05\,\text{m}^2$ filtering surface and a vacuum equivalent to an absolute pressure of $30\,\text{kN/m}^2$. The volume of filtrate collected in the first 300 s was $250\,\text{cm}^3$ and, after a further 300 s, an additional $150\,\text{cm}^3$ was collected. Assume the cake to be incompressible and the cloth resistance to be the same in the leaf as in the filter press.

Solution

In the leaf filter, filtration is at constant pressure from the start.

Thus $\qquad\qquad V^2 + 2(AL/v)V = 2(-\Delta P A^2/r\mu v)t \qquad$ (from equation 9.18)

In the filter press, a volume V_1 of filtrate is obtained under constant rate conditions in time t_1, and filtration is then carried out at constant pressure.

Thus $\qquad\qquad V_1^2 + (AL/v)V_1 = (-\Delta P A^2/r\mu v)t_1 \qquad$ (from equation 9.17)

and $\qquad (V^2 - V_1^2) + 2(AL/v)(V - V_1) = 2(-\Delta P A^2/r\mu v)(t - t_1)$

$\qquad\qquad\qquad\qquad\qquad\qquad\qquad\qquad\qquad$ (from equation 9.18)

For the leaf filter

When $\qquad\qquad\qquad t = 300\,\text{s}, \quad V = 250\,\text{cm}^3,$

and when $\qquad\qquad\qquad t = 600\,\text{s}, \quad V = 400\,\text{cm}^3$

$$A = 0{\cdot}05\,\text{m}^2, \quad \text{and} \quad -\Delta P = (101{\cdot}3 - 30{\cdot}0) = 71{\cdot}3\,\text{kN/m}^2$$

$$\therefore \qquad 250^2 + 2(0{\cdot}05L/v)\,250 = 2(71{\cdot}3 \times 0{\cdot}05^2/r\mu v)\,300$$

and $\qquad\qquad 400^2 + 2(0{\cdot}05L/v)\,400 = 2(71{\cdot}3 \times 0{\cdot}05^2/r\mu v)\,600$

i.e. $62{,}500 + 25L/v = 106{\cdot}95/r\mu v$

$160{,}000 + 40L/v = 213{\cdot}9/r\mu v$

Hence $L/v = 3500$ and $r\mu v = 7{\cdot}13 \times 10^{-4}$

For the filter press

The volume of filtrate V_1 collected during the constant rate period is given by

$[A = 2{\cdot}16\,\text{m}^2, \quad -\Delta P = (500 - 101{\cdot}3) = 398{\cdot}7\,\text{kN/m}^2, \quad t = 200\,\text{s}]:$

$$V_1^2 + 2{\cdot}16 \times 3500V_1 = (398{\cdot}7 \times 2{\cdot}16^2/7{\cdot}13 \times 10^{-4})\,200$$

i.e. $V_1^2 + 7560V_1 - 5{\cdot}218 \times 10^8 = 0$

∴ $V_1 = -3780 + \sqrt{(1{\cdot}429 \times 10^7 + 5{\cdot}218 \times 10^8)} = 1{\cdot}937 \times 10^4\,\text{cm}^3$

For the constant pressure period,

$$t - t_1 = 900\,\text{s}$$

The total volume of filtrate collected is therefore given by,

$$(V^2 - 3{\cdot}75 \times 10^8) + 15{,}120(V - 1{\cdot}937 \times 10^4) = 5{\cdot}218 \times 10^6 \times 900$$

i.e. $V^2 + 15{,}120V - 53{\cdot}64 \times 10^8 = 0$

∴ $V = -7560 + \sqrt{(5{\cdot}715 \times 10^7 + 53{\cdot}64 \times 10^8)}$

$= 6{\cdot}607 \times 10^4\,\text{cm}^3$ or $\underline{\underline{0{\cdot}066\,\text{m}^3}}$

Final rate of filtration $= -\Delta P A^2/[r\mu v(V + AL/v)]$ (from equation 9.16)

$= (398{\cdot}7 \times 2{\cdot}16^2)/[7{\cdot}13 \times 10^{-4}(6{\cdot}607 \times 10^4 + 216 \times 3500)]$

$= 35{\cdot}4\,\text{cm}^3/\text{s}$

If the viscosity of the filtrate is the same as that of the wash water,

rate of washing at $500\,\text{kN/m}^2 = 35{\cdot}4\,\text{cm}^3/\text{s}$

rate of washing at $375\,\text{kN/m}^2 = 35{\cdot}4(375 - 101{\cdot}3)/(500 - 101{\cdot}3)$

$= 24{\cdot}3\,\text{cm}^3/\text{s}$

Thus amount of wash water passing in $600\,\text{s}$

$= 600 \times 24{\cdot}3 = 1{\cdot}458 \times 10^4\,\text{cm}^3$ or $\underline{\underline{0{\cdot}0146\,\text{m}^3}}$

Problem 9.7

A sludge is filtered in a plate and frame press fitted with 50 mm frames. For the first 3600 s the slurry pump runs at maximum capacity. During this period the pressure rises to $500\,\text{kN/m}^2$ and a quarter of the total filtrate is obtained. The filtration takes a further 3600 s to complete at constant pressure and 900 s is required for emptying and resetting the press.

It is found that, if the cloths are precoated with filter aid to a depth of 1·6 mm, the cloth resistance is reduced to a quarter of its former value. What will be the increase in the overall throughput of the press if the precoat can be applied in 180 s?

Solution

The basic filtration equation is:

$$\frac{1}{A}\frac{dV}{dt} = \frac{-\Delta P}{r\mu(Vv/A+l)} \qquad \text{(equation 9.16)}$$

$$\therefore \qquad \frac{dV}{dt} = \frac{A^2-\Delta P}{r\mu v(V+Al/v)} = \frac{a}{V+b}$$

where $a = A^2 - \Delta P/r\mu v$ and $b = Al/v$.

At constant rate: $\qquad \dfrac{V_0}{t_0} = \dfrac{a}{V_0+b} \quad \text{or} \quad V_0^2 + bV_0 = at_0$ $\qquad\qquad$ (1)

At constant pressure: $\quad \frac{1}{2}(V^2 - V_0^2) + b(V - V_0) = a(t - t_0)$ $\qquad\qquad$ (2)

In case 1: $\qquad\qquad t_0 = 600\,\text{s}, \quad (t-t_0) = 3600\,\text{s}, \quad V_0 = V/4$

$\therefore$ in (1) $\qquad\qquad\qquad (V/4)^2 + bV/4 = 600a$

in (2) $\qquad\qquad\qquad \frac{1}{2}[V^2 - (V/4)^2] + b(V - V/4) = 3600a$

from which $\qquad\qquad a = 0{\cdot}000\,156V^2 \quad \text{and} \quad b = 0{\cdot}125V$

Total cycle time $= 600 + 3600 + 900 = 5100\,\text{s}$.
Filtration rate $= V/5100 = 0{\cdot}000\,196V\,\text{m}^3/\text{s}$.
For case 2, the cloth resistance is a quarter of its previous value, i.e.

$$b' = b/4$$

During the constant rate period, the slurry pump operates at constant capacity so that:

$$V_1/t_1 = V_0/t_0$$

and therefore: $\qquad\qquad \dfrac{a}{V_1 + b/4} = \dfrac{a}{V_0 + b}$

Precoat thickness $= 2 \times 1{\cdot}6 = 3{\cdot}2\,\text{mm}$.
$\therefore$ cake thickness $= 25 - 3{\cdot}2 = 21{\cdot}8\,\text{mm}$
$\therefore$ volume of cake in case 2 $= (21{\cdot}8/25) = 0{\cdot}872 \times$ volume of cake in case 1.
$\therefore$ for case 2, $\qquad\qquad$ total filtrate $V_2 = 0{\cdot}872V$

For constant pressure:

$$\frac{1}{2}[(0{\cdot}872V)^2 - V_1^2] + (b/4)(0{\cdot}872V - V_1) = a(t - t_1) \qquad (3)$$

but $\qquad\qquad V_0/t_0 = (V/4)/600 = 0{\cdot}000\,417V$

and $a = 0.000\,156V^2$

$b' = b/4 = 0.031\,25V$

∴ $0.000\,417V = \dfrac{0.000\,156V^2}{V_1^2 + 0.031\,25V}$

or $V_1 = 0.344V$

Now $\dfrac{V_1}{t_1} = \dfrac{V_0}{t_0}$ ∴ $t_1 = V_1 t_0/V_0$

but $V_0 = V/4, \quad t_0 = 600\,\text{s}, \quad V_1 = 0.344V$

∴ $t_1 = 825\,\text{s}$

Substituting $V_1 = 0.344V$ and $b' = 0.031\,25V$ into equation (3) gives:

$\frac{1}{2}[(0.872V)^2 - (0.344V)^2] + 0.031\,25V(0.872V - 0.344V) = 0.000\,156V^2(t - t_1)$

and $t - t_1 = 2159\,\text{s}$

∴   New cycle time $= 180 + 900 + 825 + 2159 = 4064\,\text{s}$

New filtration rate $= 0.872V/4064 = 0.000\,215V\,\text{m}^3/\text{s}$

∴   % increase $= (0.000\,215 - 0.000\,196)\,V \times 100/0.000\,196V$

$= \underline{\underline{9.7\%}}$

Problem 9.8

Filtration is carried out in a plate and frame filter press, with 20 frames 0.3 m square and 50 mm thick, and the rate of filtration is maintained constant for the first 300 s. During this period, the pressure is raised to $350\,\text{kN/m}^2$, and one-quarter of the total filtrate per cycle is obtained. At the end of the constant rate period, filtration is continued at a constant pressure of $350\,\text{kN/m}^2$ for a further 1800 s, after which the frames are full. The total volume of filtrate per cycle is $0.7\,\text{m}^3$ and dismantling and refitting of the press takes 500 s.

It is decided to use a rotary drum filter, 1.5 m long and 2.2 m in diameter, in place of the filter press. Assuming that the resistance of the cloth is the same in the two plants and that the filter cake is incompressible, calculate the speed of rotation of the drum which will result in the same overall rate of filtration as was obtained with the filter press. The filtration in the rotary filter is carried out at a constant pressure difference of $70\,\text{kN/m}^2$ and the filter operates with 25% of the drum submerged in the slurry at any instant.

Solution

Data from the plate and frame filter press are used to evaluate the cake and cloth resistance for use with the rotary drum filter.

For the constant rate period,

$$V_1^2 + \frac{LA}{v}V_1 = \frac{A^2 - \Delta P}{r\mu v}t_1 \qquad \text{(equation 9.17)}$$

For the subsequent constant pressure period,

$$(V^2 - V_1^2) + \frac{2LA}{v}(V - V_1) = \frac{2A^2 - \Delta P}{r\mu v}(t - t_1) \qquad \text{(equation 9.18)}$$

From the data given,

$$t_1 = 300 \text{ s}, \quad -\Delta P = 350 - 101\cdot3 = 248\cdot7 \text{ kN/m}^2,$$

$$V_1 = 0\cdot175 \text{ m}^3, \quad \text{and} \quad A = 2 \times 20 \times 0\cdot3 \times 0\cdot3 = 3\cdot6 \text{ m}^2$$

$$\therefore \quad (0\cdot175)^2 + \frac{L}{v} \times 3\cdot6 \times 0\cdot175 = \frac{(3\cdot6)^2 \times 248\cdot7 \times 10^3 \times 300}{r\mu v}$$

i.e.
$$0\cdot0306 + 0\cdot63(L/v) = 9\cdot68 \times 10^8/r\mu v \qquad \text{(i)}$$

For the constant pressure period,

$$V = 0\cdot7 \text{ m}^3, \quad V_1 = 0\cdot175 \text{ m}^3,$$

$$t - t_1 = 1800 \text{ s}, \quad A = 3\cdot6 \text{ m}^2$$

$$\therefore \quad (0\cdot7^2 - 0\cdot175^2) + 2(L/v) \times 3\cdot6(0\cdot7 - 0\cdot175) = \frac{(2 \times 3\cdot6)^2 \times 248\cdot7 \times 10^3}{r\mu v} \times 1800$$

i.e.
$$0\cdot459 + 3\cdot78(L/v) = 116\cdot08 \times 10^8/r\mu v \qquad \text{(ii)}$$

Solving (i) and (ii) simultaneously gives:

$$r\mu v = 210\cdot9 \times 10^8 \quad \text{and} \quad L/v = 0\cdot0243$$

For the rotary drum filter,

$$D = 2\cdot2 \text{ m}, \quad L = 1\cdot5 \text{ m}, \quad -\Delta P = 70 \text{ kN/m}^2$$

$$A = 2\cdot2\pi \times 1\cdot5 = 10\cdot37 \text{ m}^2$$

$$\Delta P = 70 \times 10^3 \text{ N/m}^2$$

Let θ be the time of one revolution, then as the time of filtration is $0\cdot25\theta$,

$$V^2 + 2A\frac{L}{v}V = \frac{2A^2(-\Delta P)}{r\mu v} \times 0\cdot25\theta$$

$$V^2 + 2 \times 10\cdot37 \times 0\cdot0243V = \frac{2(10\cdot37)^2 \times 70 \times 10^3 \times 0\cdot25\theta}{210\cdot9 \times 10^8}$$

$$V^2 + 0\cdot504V = 1\cdot785 \times 10^{-4}\theta$$

Now the rate of filtration $= V/t = 0\cdot7/(300 + 1800 + 500)$

$$= 2\cdot7 \times 10^{-4} \text{ m}^3/\text{s}$$

$$V = 2\cdot7 \times 10^{-4}t$$

$$\therefore \quad (2\cdot7 \times 10^{-4}t)^2 + 0\cdot504 \times 2\cdot7 \times 10^{-4}t = 1\cdot785 \times 10^{-4}t$$

from which $t = 580$ s

Hence   speed $= 1/580 = \underline{\underline{0 \cdot 002 \, \text{Hz}}}$

Problem 9.9

It is required to filter a certain slurry to produce $2 \cdot 25 \, \text{m}^3$ filtrate per working day of 8 h. The process is carried out in a plate and frame filter press with $0 \cdot 45$ m square frames and a working pressure of $450 \, \text{kN/m}^2$. The pressure is built up slowly over a period of 300 s, and during this period the rate of filtration is maintained constant.

When a sample of the slurry was filtered, using a pressure of $35 \, \text{kN/m}^2$ on a single leaf filter of filtering area $0 \cdot 05 \, \text{m}^2$, $400 \, \text{cm}^3$ of filtrate was collected in the first 300 s of filtration and a further $400 \, \text{cm}^3$ was collected during the following 600 s. Assuming that the dismantling of the filter press, the removal of the cakes and the setting up again of the press takes an overall time of 300 s, plus an additional 180 s for each cake produced, what is the minimum number of frames that need be employed? Take the resistance of the filter cloth to be the same in the laboratory tests as on the plant.

Solution

For constant pressure filtration on the leaf filter, equation 9.18 applies.

$$V^2 + 2\frac{L}{v}AV = \frac{2A^2(-\Delta P)t}{r\mu v}$$

When $t = 300$ s, $V = 0 \cdot 0004 \, \text{m}^3$, $A = 0 \cdot 05 \, \text{m}^2$, $-\Delta P = 66 \cdot 3 \, \text{kN/m}^2$

$\therefore$ $(0 \cdot 0004)^2 + 2(L/v) \times 0 \cdot 05 \times 0 \cdot 0004 = \dfrac{2 \times (0 \cdot 05)^2 \times 66 \cdot 3 \times 300}{r\mu v}$

or $1 \cdot 6 \times 10^{-7} + 4 \times 10^{-5}(L/v) = 99 \cdot 4/r\mu v$

When $t = 900$ s, $V = 800 \, \text{cm}^3 = 0 \cdot 0008 \, \text{m}^3$

and substitution gives:

$$6 \cdot 4 \times 10^{-7} + 8 \times 10^{-5}(L/v) = 298 \cdot 4/r\mu v$$

Hence $L/v = 4 \times 10^{-3}$ and $r\mu v = 3 \cdot 1 \times 10^8$

In the filter press

For the constant rate period:

$$V_1^2 + \frac{LA}{v}V_1 = \frac{A^2(-\Delta P)t_1}{r\mu v} \qquad \text{(equation 9.17)}$$

$A = 2 \times 0 \cdot 45n = 0 \cdot 9n$ where n is the number of frames

$t_1 = 300$ s

$\therefore$ $V_1^2 + 4 \times 10^{-3} \times 0 \cdot 9nV_1 = 0 \cdot 81n^2(450 - 101 \cdot 3) \times 300/3 \cdot 1 \times 10^8$

or $V_1^2 + 3 \cdot 6 \times 10^{-3}nV_1 = 2 \cdot 73 \times 10^{-4}n^2$

and $$V_1 = 0.0148n$$

For the constant pressure period:

$$\left(\frac{V^2 - V_1^2}{2}\right) + \frac{LA}{v}(V - V_1) = \frac{A^2(-\Delta P)}{r\mu v}(t - t_1)$$

Substituting for L/v, $r\mu v$, $t_1 = 300$ and $V_1 = 0.0148n$ gives:

$$\left(\frac{V^2 - 2.2 \times 10^{-4}n^2}{2}\right) + (V - 0.0148n) 4 \times 10^{-3} \times 0.9n = \frac{0.81n^2 \times 348.7}{3.1 \times 10^8}(t_f - 300)$$

or $$0.5V^2 + 1.1 \times 10^{-4}n^2 + 3.6 \times 10^{-3}nV = 9.11 \times 10^{-7}n^2 t \qquad (i)$$

Now the total cycle time $= (t_f + 300 + 180n)$ s.
Required filtration rate $= 2.25/(8 \times 3600) = 7.81 \times 10^{-5}\,\text{m}^3/\text{s}$.
Volume of filtrate $= V\,\text{m}^3$.

$$\therefore \qquad \frac{V}{t_f + 300 + 180n} = 7.81 \times 10^{-5}$$

and $$t_f = 1.28 \times 10^4 V - 300 - 180n \qquad (ii)$$

Thus value of t_f from (ii) may be substituted in equation (i) to give:

$$V^2 + V(7.2 \times 10^{-3}n - 2.34 \times 10^{-2}n^2) + (7.66 \times 10^{-4}n^2 + 3.28 \times 10^{-4}n^3) = 0 \quad (iii)$$

This equation is of the form $V^2 + AV + B = 0$ and may be solved to give:

$$V = \frac{-A \pm (A^2 - 4B)}{2}$$

where A and B are the expressions in parentheses in equation (iii). In order to find the minimum number of frames, dV/dn must be found and equated to zero. From above, $(V - a)(V - b) = 0$, where a and b are complex functions of n.
Thus $V = a$ or $V = b$ and dV/dn can be evaluated for each root.
Putting $dV/dn = 0$ gives, for the positive value,

$$n = 13$$

Problem 9.10

The relation between flow and head for a certain slurry pump can be represented approximately by a straight line, the maximum flow at zero head being $0.0015\,\text{m}^3/\text{s}$ and the maximum head at zero flow $760\,\text{m}$ of liquid.
Using this pump to feed a particular slurry to a pressure leaf filter:

(a) How long will it take to produce $1\,\text{m}^3$ of filtrate?

(b) What will be the pressure across the filter after this time?

A sample of the slurry was filtered at a constant rate of $0.00015\,\text{m}^3/\text{s}$ through a leaf filter covered with a similar filter cloth but of one-tenth the area of the full-scale unit, and after $625\,\text{s}$ the pressure across the filter was $360\,\text{m}$ of liquid. After a further $480\,\text{s}$ the pressure was $600\,\text{m}$.

Solution

For constant rate filtration through the filter leaf:

$$V^2 + \frac{LA}{v}V = \frac{A^2(-\Delta P)t}{r\mu v} \qquad \text{(equation 9.17)}$$

At a constant rate of $0{\cdot}000\,15\,\text{m}^3/\text{s}$ when time $= 625\,\text{s}$,

$$V = 0{\cdot}094\,\text{m}^3, \quad -\Delta P = 3530\,\text{kN/m}^2$$

and at $t = 1105\,\text{s}$, $\qquad V = 0{\cdot}166\,\text{m}^3 \quad \text{and} \quad -\Delta P = 5890\,\text{kN/m}^2$

Substituting these values into equation 9.17 gives:

$$(0{\cdot}094)^2 + LA/v \times 0{\cdot}094 = (A^2/r\mu v) \times 3530 \times 625$$

i.e. $\qquad\qquad 0{\cdot}0088 + 0{\cdot}094 LA/v = 2{\cdot}21 \times 10^6 A^2/r\mu v \qquad\qquad$ (i)

and $\qquad\qquad (0{\cdot}166)^2 + LA/v \times 0{\cdot}166 = (A^2/r\mu v) \times 5890 \times 1105$

i.e. $\qquad\qquad 0{\cdot}0276 + 0{\cdot}166 LA/v = 6{\cdot}51 \times 10^6 A^2/r\mu v \qquad\qquad$ (ii)

Equations (i) and (ii) may be solved simultaneously to give:

$$LA/v = 0{\cdot}0154 \quad \text{and} \quad A^2/r\mu v = 4{\cdot}64 \times 10^{-9}$$

As the full-size plant is 10 times that of the leaf filter,

$$LA/v = 0{\cdot}154 \quad \text{and} \quad A^2/r\mu v = 4{\cdot}64 \times 10^{-7}$$

If the pump develops $760\,\text{m}$ ($7460\,\text{kN/m}^2$) at zero flow and has zero head at $Q = 0{\cdot}0015\,\text{m}^3/\text{s}$, its performance can be expressed as:

$$\Delta P = 7460 - (7460/0{\cdot}0015)Q$$

or $\qquad\qquad\qquad \Delta P = 7460 - 4{\cdot}97 \times 10^6 Q\,(\text{kN/m}^2)$

Now $\qquad\qquad\qquad \dfrac{dV}{dt} = \dfrac{A^2(-\Delta P)}{r\mu v(V + LA/v)} \qquad\qquad$ (equation 9.16)

Substituting for $-\Delta P$ and the filtration constants gives:

$$\frac{dV}{dt} = \frac{A^2}{r\mu v}\frac{(7460 - 4{\cdot}97 \times 10^6\,dV/dt)}{(V + 0{\cdot}154)}$$

Since $Q = dV/dt$:

$$\frac{dV}{dt} = \frac{4{\cdot}67 \times 10^{-7}[7460 - 4{\cdot}97 \times 10^6\,(dV/dt)]}{(V + 0{\cdot}154)}$$

$$(V + 0{\cdot}154)\,dV = 3{\cdot}46 \times 10^{-3} - 2{\cdot}31\,dV/dt$$

The time to collect $1\,\text{m}^3$ is then given by:

$$\int_0^1 (V + 0{\cdot}154 + 2{\cdot}31)\,dV = \int_0^t (3{\cdot}46 \times 10^{-3})\,dt$$

and $$t = 857\,\mathrm{s}$$

The pressure at this time is found by substitution in equation 9.17 with $V = 1\,\mathrm{m}^3$ and $t = 857\,\mathrm{s}$:

$$1^2 + 0\cdot154 \times 1 = 4\cdot64 \times 10^{-7} \times 857\Delta P$$

and $$-\Delta P = 2902\,\mathrm{kN/m}^2$$

Problem 9.11

A slurry containing 40% by weight solid is to be filtered on a rotary drum filter 2 m diameter and 2 m long which normally operates with 40% of its surface immersed in the slurry and under a pressure of $17\,\mathrm{kN/m}^2$. A laboratory test on a sample of the slurry using a leaf filter of area 200 cm² and covered with a similar cloth to that on the drum produced 300 cm³ of filtrate in the first 60 s and 140 cm³ in the next 60 s, when the leaf was under an absolute pressure of $17\,\mathrm{kN/m}^2$. The bulk density of the dry cake was $1500\,\mathrm{kg/m}^3$ and the density of the filtrate $1000\,\mathrm{kg/m}^3$. The minimum thickness of cake which could be readily removed from the cloth was 5 mm.

At what speed should the drum rotate for maximum throughput and what is this throughput in terms of the weight of the slurry fed to the unit per unit time?

Solution

For the leaf filter:

$$A = 0\cdot02\,\mathrm{m}^2, \quad \Delta P = 101\cdot3 - 17 = 84\cdot3\,\mathrm{kN/m}^2 = 84{,}300\,\mathrm{N/m}^2$$

When $t = 60$, $$V = 0\cdot0003\,\mathrm{m}^3$$

When $t = 120$, $$V = 0\cdot000\,44\,\mathrm{m}^3$$

These figures are substituted into the constant pressure filtration equation 9.18.

$$V^2 + \frac{2LAV}{v} = \frac{2(-\Delta P)A^2 t}{r\mu v} \qquad \text{(equation 9.18)}$$

This enables the filtration constants to be determined as:

$$L/v = 2\cdot19 \times 10^{-3} \quad \text{and} \quad r\mu v = 3\cdot48 \times 10^{10}$$

For the rotary filter equation 9.18 applies as the whole operation is at constant pressure. The maximum throughput will be obtained when the cake thickness is a minimum, i.e. 5 mm = 0·005 m.

Area of filtering surface $= 2\pi \times 2 = 4\pi\,\mathrm{m}^2$.

Bulk volume of cake deposited $= 4\pi \times 0\cdot005 = 0\cdot063\,\mathrm{m}^3/\mathrm{rev}$.

Let the rate of filtrate production $= w\,\mathrm{kg/s} = 0\cdot001 w\,\mathrm{m}^3/\mathrm{s}$.

For a 40% slurry: $$S/(S + w) = 0\cdot4$$

and weight solids $= 0\cdot66 w$.

∴   volume of solids deposited/s $= 0.66w/1500 = 4.4 \times 10^{-4}w\,\text{m}^3/\text{s}$

If one revolution takes t s,
$$4.4 \times 10^{-4}wt = 0.063$$
and weight $= 143\,\text{kg}$.

Rate of production of filtrate $= 0.001w\,\text{m}^3/\text{s} = V/t$

∴
$$V^2 = 1 \times 10^{-6}w^2t^2$$
$$= 1 \times 10^{-6}(143)^2$$
$$= 0.02\,\text{m}^6$$
$$V = 0.141\,\text{m}^3$$

Substituting $V = 0.141\,\text{m}^3$ and the constants into equation 9.18,
$$(0.141)^2 + 2 \times 2.19 \times 10^{-3} \times 0.141 = 2 \times 84,300 \times (4\pi)^2t/3.48 \times 10^{10}$$
from which $t = 26.95$ s $=$ time of submergence/rev.

∴   time for 1 rev $= 26.9/0.4 = 67.3$ s

∴   speed $= 1/67.3 = \underline{\underline{0.015\,\text{Hz}}}$ (0.9 rev/min)

$$w = 143/67.3 = 2.11\,\text{kg/s}$$
$$S = 0.66 \times 2.11\,\text{kg/s}$$

∴   weight of slurry $= 1.66 \times 2.11 = \underline{\underline{3.5\,\text{kg/s}}}$

Problem 9.12

A continuous rotary filter is required for an industrial process for the filtration of a suspension to produce $0.002\,\text{m}^3/\text{s}$ of filtrate. A sample was tested on a small laboratory filter of area $0.023\,\text{m}^2$ to which it was fed by means of a slurry pump to give filtrate at a constant rate of $12.5\,\text{cm}^3/\text{s}$. The pressure difference across the test filter increased from $14\,\text{kN/m}^2$ after 300 s filtration to $28\,\text{kN/m}^2$ after 500 s at which time the cake thickness had reached 38 mm. Suggest suitable dimensions and operating conditions for the rotary filter, assuming that the resistance of the cloth used is one-half that on the test filter, and that the vacuum system is capable of maintaining a constant pressure difference of $70\,\text{kN/m}^2$ across the filter.

Solution

Data from the laboratory filter may be used to find the cloth and cake resistance of the rotary filter. For the laboratory filter operating under constant rate conditions:
$$V_1^2 + \frac{LA}{v}V_1 = \frac{A^2(-\Delta P)t}{r\mu v} \qquad \text{(equation 9.17)}$$

$A = 0.023\,\text{m}^2$ and the filtration rate $= 12.5\,\text{cm}^3/\text{s}$.

At $t = 300\,\text{s}$,

$$-\Delta P = 14\,\text{kN/m}^2 \quad \text{and} \quad V_1 = 3750\,\text{cm}^3 = 3.75 \times 10^{-3}\,\text{m}^3$$

When $t = 900\,\text{s}$,

$$-\Delta P = 28\,\text{kN/m}^2 \quad \text{and} \quad V_1 = 11{,}250\,\text{cm}^3 = 1.125 \times 10^{-2}\,\text{m}^3$$

Hence $\quad (3.75 \times 10^{-3})^2 + (L/v) \times 0.023 \times 3.75 \times 10^{-3} = \dfrac{14}{r\mu v} \times (0.023)^2 \times 300$

$$1.41 \times 10^{-5} + 8.63 \times 10^{-5}(L/v) = 2.22/r\mu v$$

and $\quad (1.25 \times 10^{-2})^2 + (L/v) \times 0.023 \times 1.125 \times 10^{-3} = \dfrac{28 \times (0.023)^2}{r\mu v} \times 900$

$$1.27 \times 10^{-4} + 2.59 \times 10^{-4}(L/v) = 13.33/r\mu v$$

from which $\qquad L/v = 0.164; \quad r\mu v = 7.86 \times 10^4$

If the cloth resistance is halved by using the rotary filter, $L/v = 0.082$. As the filter operates at constant pressure, equation 9.18 is applicable:

$$V^2 \frac{2LA}{v} = \frac{2A^2(-\Delta P)t}{r\mu v} \qquad \text{(equation 9.18)}$$

Let $\theta = $ time for 1 rev $\times$ fraction submerged and $V' = $ volume of filtrate/rev (given by equation 9.18).

Assume speed $= 0.0167\,\text{Hz}$ (1 rev/min) and 20% submergence.

Then $\qquad\qquad\qquad\qquad \theta = 60 \times 0.2 = 12\,\text{s}$

$\therefore \qquad\qquad\qquad V'^2 + 2 \times 0.082AV' = \dfrac{2A^2 \times 70 \times 12}{7.86 \times 10^4}$

or $\qquad\qquad\qquad V'^2 + 0.164AV' = 0.0214A^2$

from which $\qquad\qquad\qquad A/V' = 11.7$

The required rate of filtration $= 0.002\,\text{m}^3/\text{s}$.

$\therefore \qquad\qquad V' = \text{volume/rev} = 0.002 \times 60 = 0.12\,\text{m}^3$

$\therefore \qquad\qquad A = 11.7 \times 0.12 = 1.41\,\text{m}^2$

If $L = D$, $\qquad\qquad$ area of drum $= \pi D^2 = 1.41\,\text{m}^2$

and $\qquad\qquad\qquad D = L = \underline{\underline{0.67\,\text{m}}}$

The cake thickness on the drum should now be checked.

$$v = AL/V$$

and from data on the laboratory filter,

$$v = 0.023 \times 0.038/1.125 \times 10^{-2}$$

$$= 0.078$$

Hence, cake thickness on drum $= vV'/A = 0\cdot078/11\cdot7$

$$= 0\cdot0067 \text{ m} = \underline{6\cdot7 \text{ mm}} \quad \text{which is acceptable}$$

Problem 9.13

A rotary drum filter 1·2 m diameter and 1·2 m long can handle 6·0 kg/s of slurry containing 10% of solids when rotated at 0·005 Hz. By increasing the speed to 0·008 Hz, it is found that it can handle 7·2 kg/s. What will be the percentage change in the amount of wash water which can be applied to each kilogram of cake caused by this increase of speed? What are the limitations to increased production by increase in the speed of rotation of the drum, and what is the theoretical maximum quantity of slurry which can be handled?

Solution

For constant pressure filtration:

$$\frac{dV}{dt} = \frac{A^2(-\Delta P)}{r\mu v[V+(LA/v)]} = \frac{a}{V+b} \quad \text{(say)}$$

$$\therefore \qquad V^2/2 + bV = at$$

For case 1: 1 rev. takes $1/0\cdot005 = 200$ s and the rate $= V_1/200$.
For case 2: 1 rev. takes $1/0\cdot008 = 125$ s and the rate $= V_2/125$.

But

$$\frac{V_1/200}{V_2/125} = \frac{6\cdot0}{7\cdot2}$$

$$\therefore \qquad V_1/V_2 = 1\cdot33 \quad \text{and} \quad V_2 = 0\cdot75V_1$$

For case 1, using the filtration equation,

$$V_1^2 + 2bV_1 = 2a \times 200$$

and for case 2,

$$V_2^2 + 2bV_2 = 2a \times 125$$

Substituting $V_2 = 0\cdot75V_1$ in these two equations allows the filtration constants to be found as

$$a = 0\cdot003\,75V_1^2 \quad \text{and} \quad b = 0\cdot25V_1$$

The rate of flow of wash water will equal the final rate of filtration so that for case 1:

Wash water rate $= a/(V_1 + b)$.

Wash water per rev. $\propto 200a/(V_1 + b)$.

Wash water/rev. per unit solids $\propto 200a/V_1(V_1 + b)$,

i.e.
$$\propto (200 \times 0\cdot003\,75V_1^2)/V_1(V_1 + 0\cdot25V_1)$$

$$\propto 0\cdot6.$$

Similarly for case 2, the wash water per rev. per unit solids is proportional to:

$$125a/V^2(V_2+b)$$

$$\propto (125 \times 0.003\ 75V_1^2)/0.75V_1(0.75V_1 + 0.25V_1)$$

$$\propto 0.625$$

Hence the % increase $= [(0.625 - 0.6)/0.6] \times 100$

$$= \underline{\underline{4.17\%}}$$

As $0.5V^2 + bV = at$, the rate of filtration V/t is given by:

$$a/(0.5V+b)$$

The highest rate will be achieved when V tends to zero and $(V/t)_{max} = a/b$

$$= 0.003\ 75V_1^2/0.25V_1$$

$$= 0.015V_1$$

For case 1, the rate $= V_1/200 = 0.005V_1$.

Hence the limiting rate is three times the original rate,

i.e. $\qquad\qquad\qquad\qquad \underline{\underline{18.0\,\text{kg/s}}}$

Problem 9.14

When an aqueous slurry is filtered in a plate and frame press, fitted with two 50 mm thick frames each 150 mm square at $450\,\text{kN/m}^2$, the frames are filled in 3·5 ks. The liquid in the slurry has the same density as water. The slurry is then filtered in a perforate basket centrifuge with a basket 300 mm diameter and 200 mm deep. If the radius of the inner surface of the slurry is maintained constant at 75 mm and the speed of rotation is 65 Hz, how long will it take to produce as much filtrate as was obtained from a single cycle of operations with the filter press?

Assume that the filter cake is incompressible and that the resistance of the cloth is equivalent to 3 mm of cake in both cases.

Solution

In the filter press

$$V^2 + 2(AL/v)V = 2(-\Delta P)A^2/r\mu v)t \quad (\because \quad V = 0 \quad \text{when} \quad t = 0)$$

$$\text{(from equation 9.18)}$$

Now $\qquad\qquad\qquad\qquad V = lA/v$

$$\therefore \qquad\qquad l^2A^2/v^2 + 2(AL/v)(lA/v) = 2(-\Delta PA^2/r\mu v)t$$

$$\therefore \qquad\qquad l^2 + 2Ll = 2(-\Delta Pv/r\mu)t$$

For one cycle:

$$l = 25\,mm = 0.025\,m, \quad L = 3\,mm = 0.003\,m$$

$$\Delta P = (450 - 101.3) = 348.7\,kN/m^2 \quad or \quad 3.49 \times 10^5\,N/m^2$$

$$t = 3500\,s$$

$$\therefore \qquad 0.025^2 + (2 \times 0.003 \times 0.025) = 2 \times 3.49 \times 10^5 \times 3500(v/r\mu)$$

$$\therefore \qquad r\mu/v = 3.15 \times 10^{12}$$

In the centrifuge

$$(b^2 - b'^2)(1 + 2L/b) + 2b'^2 \ln(b'/b) = (2vt\rho\omega^2/r\mu)(b^2 - x^2) \quad \text{(equation 9.51)}$$

$$b = 0.15\,m, \quad H = 0.20\,m$$

Volume of cake $= 2 \times 0.050 \times 0.15^2 = 0.002\,25\,m^3$.

$$\therefore \qquad \pi(b^2 - b'^2) \times 0.20 = 0.002\,25$$

$$\therefore \qquad (b^2 - b'^2) = 0.003\,58$$

$$\therefore \qquad b'^2 = 0.15^2 - 0.003\,58 = 0.0189\,m^2$$

$$\therefore \qquad b' = 0.138\,m$$

$$x = 75\,mm = 0.075\,m, \quad w = 65 \times 2\pi = 408.4\,rad/s$$

The time taken to produce the same volume of filtrate or cake as in one cycle of the filter press is therefore given by:

$$(0.15^2 - 0.138^2)(1 + 2 \times 0.003/0.15) + 2(0.0189) \ln (0.138/0.15)$$

$$= [2t \times 1000 \times 408.4^2/(3.15 \times 10^{12})](0.15^2 - 0.075^2)$$

$$0.003\,59 - 0.003\,15 = 1.787 \times 10^{-6}t$$

$$t = 4.4 \times 10^{-4}/1.787 \times 10^{-6}$$

$$= \underline{\underline{246\,s}} \quad (\simeq 4\,min)$$

Problem 9.15

A rotary drum filter of area $3\,m^2$ operates with an internal pressure of $30\,kN/m^2$ and with 30% of its surface submerged in the slurry. Calculate the rate of production of filtrate and the thickness of cake when it rotates at $0.0083\,Hz$ ($0.5\,rev/min$), if the filter cake is incompressible and the filter cloth has a resistance equal to that of 1 mm of cake.

It is desired to increase the rate of filtration by raising the speed of rotation of the drum. If the thinnest cake that can be removed from the drum has a thickness of 5 mm, what is the maximum rate of filtration which can be achieved and what speed of rotation of the drum is required?

Voidage of cake $= 0.4$.

Specific resistance of cake $= 2 \times 10^{12}/m^2$.

Density of solids = 2000 kg/m^3.

Density of filtrate = 1000 kg/m^3.

Viscosity of filtrate = 10^{-3} N s/m^2.

Slurry concentration = 20% by weight solids.

Solution

A 20% slurry contains 20 kg solids/80 kg solution.

Volume of cake = 20/2000(1 − 0·4) = 0·0167 m^3.

Volume of liquid in cake = 0·167 × 0·4 = 0·0067 m^3.

Volume of filtrate = (80/1000) − 0·0067 = 0·0733 m^3.

∴
$$v = 0.0167/0.0733 = 0.23$$

The rate of filtration is given by:

$$\frac{dV}{dt} = \frac{A^2(-\Delta P)}{r\mu v[V+(LA/v)]} \qquad \text{(equation 9.16)}$$

In this problem:

$$A = 3\,\text{m}^2$$
$$\Delta P = 101.3 - 30 = 71.3\,\text{kN/m}^2 = 71.3 \times 10^3\,\text{N/m}^2$$
$$r = 2 \times 10^{12}/\text{m}^2$$
$$\mu = 1 \times 10^{-3}\,\text{N s/m}^2$$
$$v = 0.23$$
$$L = 1\,\text{mm} = 1 \times 10^{-3}\,\text{m}$$

∴
$$\frac{dV}{dt} = \frac{3^2 \times 71.3 \times 10^3}{0.23 \times 2 \times 10^{12} \times 1 \times 10^{-3}[V+(1 \times 10^{-3} \times 3/0.23)]}$$

$$= \frac{1.395 \times 10^3}{V + 0.013}$$

From which
$$V^2/2 + 0.013V = 1.395 \times 10^{-3}t$$

If the speed = 0·0083 Hz, 1 revolution takes 120·5 s and a given piece is immersed for 120·5 × 0·3 = 36·2 s. When t = 36·2 s, V may be found by substitution to be 0·303 m^3. Hence the rate of filtration = 0·303/120·5 = 0·0025 m^3/s.

Volume of filtrate for 1 rev = 0·303 m^3.

Volume of cake = 0·23 × 0·303 = 0·07 m^3.

∴
$$\text{cake thickness} = 0.07/3 = 0.023\,\text{m} = \underline{\underline{23\,\text{mm}}}$$

As the thinnest cake = 5 mm, volume of cake = 3 × 0·005 = 0·015 m^3.

As $v = 0.23$, volume of filtrate $= 3 \times 0.005/0.23 = 0.065 \, \text{m}^3$.

$\therefore$
$$(0.065)^2/2 + 0.013 \times 0.065 = 1.395 \times 10^{-3}t$$

and
$$t = 2.12 \, \text{s}$$

$\therefore$ time for 1 rev $= 2.12/0.3 = 7.1 \, \text{s}$ and   speed $= \underline{0.14 \, \text{Hz}}$ (8.5 rev/min)

Maximum filtrate rate $= 0.065 \, \text{m}^3$ in $7.1 \, \text{s}$

$$= 0.065/7.1 = \underline{\underline{0.009 \, \text{m}^3/\text{s}}}$$

LEACHING

Problem 10.1

0·4 kg/s of dry seashore sand, containing 1% by weight of salt, is to be washed with 0·4 kg/s of fresh water running countercurrent to the sand through two classifiers in series. Assume that perfect mixing of the sand and water occurs in each classifier and that the sand discharged from each classifier contains one part of water for every two of sand (by weight).

If the washed sand is dried in a kiln drier, what percentage of salt will it retain?

What wash rate would be required in a single classifier in order to wash the sand equally well?

Solution

The problem entails a mass balance around the two stages. Let x kg/s salt be in the underflow discharge from stage 1.

Salt in feed to stage 2 = $(0{\cdot}4 \times 1/100) = 0{\cdot}004$ kg/s.

The sand passes through each stage and hence sand in underflow from stage 1 = 0·4 kg/s, which is associated with $(0{\cdot}4/2) = 0{\cdot}2$ kg/s water (assuming constant underflow). Similarly, 0·2 kg/s water enters stage 1 in the underflow and 0·4 kg/s enters in the overflow. Making a water balance around stage 1, water in the overflow discharge = 0·4 kg/s.

In the underflow discharge from stage 1, x kg/s salt is associated with 0·2 kg/s water, and hence the salt associated with the 0·4 kg/s water in the overflow discharge = $(x \times 0{\cdot}4/0{\cdot}2) = 2x$ kg/s. This assumes that the overflow and underflow solutions have the same concentration.

Considering stage 2, 0·4 kg/s water enters in the overflow and 0·2 kg/s leaves in the underflow.

∴ water in overflow from stage 2 = $(0{\cdot}4 - 0{\cdot}2) = 0{\cdot}2$ kg/s.

The salt entering is 0·004 kg/s in the underflow and $2x$ in the overflow—a total of $(0{\cdot}004 + 2x)$ kg/s. The exit underflow and overflow concentrations must be the same, and hence the salt associated with 0·2 kg/s water in each stream is

$$(0{\cdot}004 + 2x)/2 = (0{\cdot}002 + x)\ \text{kg/s}$$

Making an overall salt balance,

$$0{\cdot}004 = x + (0{\cdot}002 + x)$$

∴

$$x = 0{\cdot}001\ \text{kg/s}$$

This is associated with 0·4 kg/s sand and hence

$$\text{percentage salt in dried sand} = (0{\cdot}001 \times 100)/(0{\cdot}4 + 0{\cdot}001)$$

$$= 0{\cdot}249\%$$

The same result may be obtained by applying equation 10.16 over the washing stage 1:

$$S_{n+1}/S_1 = (R-1)/(R^{n+1}-1)$$

In this case: $R = 0{\cdot}4/0{\cdot}2 = 2$,

$$n = 1, \quad S_2 = x,$$

$$S_1 = (0{\cdot}002 + x),$$

$$\therefore \qquad x/(0{\cdot}002 + x) = (2-1)/(2^2 - 1) = 0{\cdot}33$$

$$x = 0{\cdot}000\,667/0{\cdot}667 = 0{\cdot}001\ \text{kg/s}$$

and the percentage salt in the sand = 0·249% as before.

Considering a *single stage*,

Let y kg/s be the overflow feed of water. Since 0·2 kg/s water leaves in the underflow, the water in the overflow discharge $= (y - 0{\cdot}2)$ kg/s. With a feed of 0·004 kg/s salt and 0·001 kg/s salt in the underflow discharge, the salt in the overflow discharge $= 0{\cdot}003$ kg/s. The ratio of salt/solution must be the same in both discharge streams or:

$$(0{\cdot}001)/(0{\cdot}20 + 0{\cdot}001) = 0{\cdot}003/(0{\cdot}003 + y - 0{\cdot}2)$$

$$\therefore \qquad y = 0{\cdot}8\ \text{kg/s}$$

Problem 10.2

Caustic soda is manufactured by the lime-soda process according to the following equation:

$$Na_2CO_3 + Ca(OH)_2 = 2NaOH + CaCO_3$$

A solution of sodium carbonate in water (0·25 kg/s Na_2CO_3) is treated with the theoretical requirement of lime and, after the reaction is complete, the $CaCO_3$ sludge, containing by weight 1 part of $CaCO_3$ per 9 parts of water, is fed continuously to three thickeners in series and is washed countercurrently. Calculate the necessary rate of feed of neutral water to the thickeners, so that the calcium carbonate, on drying, contains only 1% of sodium hydroxide. The solid discharged from each thickener contains 1 part by weight of calcium carbonate to 3 of water.

Solution

$$Na_2CO_3 + Ca(OH)_2 = 2NaOH + CaCO_3$$

$$106\ \text{kg} = 80\ \text{kg} + 100\ \text{kg}$$

Let x_1', x_2', x_3' be the solute/solvent ratios in thickeners 1, 2, and 3. The quantities of

$CaCO_3$, NaOH, and water in each of the streams will be calculated for every 100 kg of calcium carbonate.

	$CaCO_3$	NaOH	Water
Overall balance			
Feed from reactor	100	80	900
Feed as wash water	—	—	W_f (say)
Product-underflow	100	$300x'_3$	300
Product-overflow	—	$80 - 300x'_3$	$600 + W_f$
Thickener 1			
Feed from reactor	100	80	900
Feed-overflow	—	$300(x'_1 - x'_3)$	W_f
Product-underflow	100	$300x'_1$	300
Product-overflow	—	$80 - 300x'_3$	$600 + W_f$
Thickener 2			
Feed-underflow	100	$300x'_1$	300
Feed-overflow	—	$300(x'_2 - x'_3)$	W_f
Product-underflow	100	$300x'_2$	300
Product-overflow	—	$300(x'_1 - x'_3)$	W_f
Thickener 3			
Feed-underflow	100	$300x'_2$	300
Feed-water	—	—	W_f
Product-underflow	100	$300x'_3$	300
Product-overflow	—	$300(x'_2 - x'_3)$	W_f

Since the final underflow must contain only 1% of NaOH,

$$300x'_3/100 = 0.01$$

If equilibrium is achieved in each of the thickeners, the ratio of NaOH to water will be the same in the underflow and the overflow.

Thus
$$300(x'_2 - x'_3)/W_f = x'_3$$
$$300(x'_1 - x'_3)/W_f = x'_2$$
$$(80 - 300x'_3)/(600 + W_f) = x'_1$$

Simultaneous solution of these four equations gives:

$$x'_1 = 0.05, \quad x'_2 = 0.0142, \quad x'_3 = 0.0033, \quad W_f = 980$$

Thus the amount of water required for washing 100 kg/s $CaCO_3$ is 980 kg/s.
Solution fed to the reactor contains 25 kg/s Na_2CO_3. This is equivalent to 23.6 kg/s of $CaCO_3$.
Thus actual feed of water required $= (980 \times 23.6/100)$

$$= \underline{\underline{230\,kg/s}}$$

Problem 10.3

How many stages are required for 98% extraction of a material containing 18% of extractable matter (having a specific gravity of 2.7) and which requires 200 volumes of liquid per 100 volumes of solid for it to be capable of being pumped to the next stage? . The strong solution is to have a concentration of 100 kg/m³.

Solution

Take as a basis *100 kg solids fed to the plant.*

This contains 18 kg solute and 82 kg inert material. The extraction is 98% and hence $(0.98 \times 18) = 17.64$ kg solute appears in the liquid product, leaving $(18 - 17.64) = 0.36$ kg solute in the washed solids. The concentration of the liquid product is 100 kg/m^3 and hence the volume of the liquid product $= (17.64/100) = 0.1764 \text{ m}^3$.

Volume of solute in liquid product $= (17.64/2700) = 0.006\,53 \text{ m}^3$.

Volume of solvent in liquid product $= (0.1764 - 0.006\,53) = 0.1699 \text{ m}^3$.

Mass of solvent in liquid product $= 0.1699\rho$ kg where $\rho \text{ kg/m}^3$ is the density of the solvent.

In the washed solids, total solids $= 82$ kg or $(82/2700) = 0.0304 \text{ m}^3$.

Volume of solution in the washed solids $= (0.0304 \times 200/100) = 0.0608 \text{ m}^3$.

Volume of solute in solution $= (0.36/2700) = 0.0001 \text{ m}^3$.

Volume of solvent in washed solids $= (0.0608 - 0.0001) = 0.0607 \text{ m}^3$.

∴   mass of solvent in washed solids $= 0.0607\rho$ kg

and   mass of solvent fed to the plant $= (0.0607 + 0.1699)\rho = 0.2306\rho$ kg

The *overall balance* is therefore:

	Inerts	Solute	Solvent
Feed to plant	82	18	—
Wash liquor	—	—	0.2306ρ
Washed solids	82	0.36	0.0607ρ
Liquid product	—	17.64	0.1699ρ

$$\frac{\text{Solvent discharged in the overflow}}{\text{Solvent discharged in the underflow}}, \quad R = 0.2306\rho/0.0607\rho = 3.80$$

The overflow product contains $100 \text{ kg solute/m}^3$ solution. This concentration is the same as the underflow from the first thickener and hence the material fed to the washing thickeners contains 82 kg inerts and 0.0608 m^3 solution containing $(100 \times 0.0608) = 6.08$ kg solute.

Thus, in equation 10.16:

$$(3.80 - 1)/(3.80^{n+1} - 1) = 0.36/6.08$$

$$3.80^{n+1} = 48.28$$

and $n = 1.89$, say 2 washing thickeners.

Thus a total of 3 thickeners is required.

Problem 10.4

Soda ash is mixed with lime and the liquor from the second of three thickeners and passes to the first thickener where separation is effected. The quantity of this caustic solution leaving the first thickener is such as to yield 10 Mg of caustic soda per day of

24 h. The solution contains 95 kg of caustic soda per 1000 kg of water, whilst the sludge leaving each of the thickeners consists of one part of solids to one of liquid.

Determine: (a) the weight of solids in the sludge;

(b) the weight of water admitted to the third thickener;

(c) the percentages of caustic soda in the sludges leaving the respective thickeners.

Solution

Basis: 100 Mg $CaCO_3$ in the sludge leaving each thickener.

Let x_1, x_2, x_3 be the ratio of caustic soda to solution by mass in each thickener.

In order to produce 100 Mg $CaCO_3$, 106 Mg Na_2CO_3 must react giving 80 Mg NaOH by the equation:

$$Na_2CO_3 + Ca(OH)_2 = 2NaOH + CaCO_3$$

For the purposes of calculation it may be assumed that a mixture of 100 Mg $CaCO_3$ and 80 Mg NaOH is fed to the first thickener and w Mg water as the overflow feed to the third thickener. The mass balances are then made as follows:

	$CaCO_3$	NaOH	Water
Overall			
Underflow feed	100	80	—
Overflow feed	—	—	w
Underflow product	100	$100x_3$	$100(1-x_3)$
Overflow product	—	$(80-100w_3)$	$w-100(1-x_3)$
Thickener 1			
Underflow feed	100	80	—
Overflow feed	—	$100(x_1-x_3)$	$w+100(x_3-x_1)$
Underflow product	100	$100x_1$	$100(1-x_1)$
Overflow product	—	$80-100x_3$	$w-100(1-x_3)$
Thickener 2			
Underflow feed	100	$100x_1$	$100(1-x_1)$
Overflow feed	—	$100(x_2-x_3)$	$w+100(x_3-x_2)$
Underflow product	100	$100x_2$	$100(1-x_2)$
Overflow product	—	$100(x_1-x_3)$	$w+100(x_3-x_1)$
Thickener 3			
Underflow feed	100	$100x_2$	$100(1-x_2)$
Overflow feed	—	—	w
Underflow product	100	$100x_3$	$100(1-x_3)$
Overflow product	—	$100(x_2-x_3)$	$w+100(x_3-x_2)$

In the overflow product, 0·095 Mg NaOH is associated with 1 Mg water.

$$\therefore \quad x_1 = 0·095/(1+0·095) = 0·0868 \text{ Mg/Mg solution} \quad (i)$$

Assuming equilibrium is attained in each thickener, the concentration of NaOH in the overflow product is equal to the concentration of NaOH in the solution in the underflow product.

$$\therefore \quad x_3 = [100(x_2-x_3)]/[100(x_2-x_3) + w - 100(x_2-x_3)] = 100(x_2-x_3)/w \quad (ii)$$

$$x_2 = [100(x_1-x_3)]/[100(x_1-x_3) + w - 100(x_1-x_3)] = 100(x_1-x_3)/w \quad (iii)$$

$$x_3 = (80-100x_3)/[80 - 100x_3 + w - 100(1-x_3)] = (80-100x_3)/(w-20) \quad (iv)$$

and solving equations (i)–(iv) simultaneously,

$$x_3 = 0.0010 \, \text{Mg/Mg}, \quad x_2 = 0.0093 \, \text{Mg/Mg}, \quad x_1 = 0.0868 \, \text{Mg/Mg}$$

and $w = 940.5 \, \text{Mg/100 Mg CaCO}_3$

The overflow product $= w - 100(1 - x_3) = 840.6 \, \text{Mg/100 Mg CaCO}_3$.
Actual flow of caustic solution $= (10/0.0868) = 115 \, \text{Mg/day}$.

$\therefore$ mass of $CaCO_3$ in sludge $= (100 \times 115/840.6) = \underline{13.7 \, \text{Mg/day}}$

mass of water fed to third thickener $= 940.5 \, \text{Mg/100 Mg CaCO}_3$

$$= (940.5 \times 13.7/100) = \underline{129 \, \text{Mg/day}}$$

The total mass of sludge leaving each thickener $= 200 \, \text{Mg/100 Mg CaCO}_3$.
Mass of caustic soda in the sludge $= 100x_1 \, \text{Mg/100 Mg CaCO}_3$ and hence concentration of caustic in sludge leaving:

thickener 1 $= (100 \times 0.0868 \times 100/200) = \underline{4.34\%}$

thickener 2 $= (100 \times 0.0093 \times 100/200) = \underline{0.47\%}$

thickener 3 $= (100 \times 0.0010 \times 100/200) = \underline{0.05\%}$

Problem 10.5

Seeds, containing 20% by weight of oil, are extracted in a countercurrent plant, and 90% of the oil is recovered in a solution containing 50% by weight of oil. If the seeds are extracted with fresh solvent and 1 kg of solution is removed in the underflow in association with every 2 kg of insoluble matter, how many ideal stages are required?

Solution

This problem will be solved using the graphical method.
Since the seeds contain 20% of oil,

$$x_{A_1} = 0.2 \quad \text{and} \quad x_{B_1} = 0.8$$

The final solution contains 50% of oil.

Thus $y_{A_1} = 0.5 \quad \text{and} \quad y_{S_1} = 0.5$

The solvent which is used for extraction is pure and therefore

$$y_{S \cdot n+1} = 1$$

Now 1 kg of insoluble solid in the washed product is associated with 0.5 kg of solution and 0.025 kg of oil. Thus

$$x_{A \cdot n+1} = 0.0167, \quad x_{B \cdot n+1} = 0.6667 \quad \text{and} \quad x_{S \cdot n+1} = 0.3166$$

The mass fraction of insoluble material in the underflow is constant and equal to 0·667. The composition of the underflow is therefore represented, on the diagram, by a straight line parallel to the hypotenuse of the triangle with an intercept of 0·333 on the two main axes.

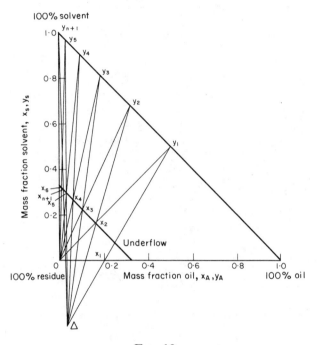

FIG. 10a

The difference point is now found by drawing in the two lines connecting x_1 and y_1 and x_{n+1} and y_{n+1}.

The graphical construction described in the text is then used and it is seen from the diagram (Fig. 10a) that x_{n+1} lies in between x_5 and x_6.

Thus <u>5 thickeners</u> are adequate and produce the required degree of extraction.

Problem 10.6

It is desired to recover precipitated chalk from the causticising of soda ash. After decanting the liquor from the precipitators the sludge has the composition 5% $CaCO_3$, 0·1% NaOH, and the balance water.

1000 Mg/day of this sludge is fed to two thickeners where it is washed with 200 Mg/day of neutral water. The pulp removed from the bottom of the thickeners contains 4 kg of water per kg of chalk. The pulp from the last thickener is taken to a rotary filter and concentrated to 50% solids and the filtrate is returned to the system as wash water.

Calculate the net percentage of $CaCO_3$ in the product after drying.

Solution

Basis: 1000 Mg/day sludge fed to the plant

Let x_1 and x_2 be the solute/solvent ratio in thickeners 1 and 2. The mass balances become:

	$CaCO_3$	NaOH	Water
Overall			
Underflow feed	50	1	949
Overflow feed	—	—	200
Underflow product	50	$200x_2$	200
Overflow product	—	$(1 - 200x_2)$	949
Thickener 1			
Underflow feed	50	1	949
Overflow feed	—	$200(x_1 - x_2)$	200
Underflow product	50	$200x_1$	200
Overflow product	—	$(1 - 200x_2)$	949
Thickener 2			
Underflow feed	50	$200x_1$	200
Overflow feed	—	—	200
Underflow product	50	$200x_2$	200
Overflow product	—	$200(x_1 - x_2)$	200

The solute/solvent ratio will be the same in the overflow and underflow products of each thickener assuming equilibrium is attained.

$$\therefore \qquad x_2 = 200(x_1 - x_2)/200 \quad \text{or} \quad x_2 = 0.5x_1$$

and

$$x_1 = (1 - 200x_2)/949$$

$$\therefore \qquad x_1 = 0.000\,954 \quad \text{and} \quad x_2 = 0.000\,477$$

The underflow product contains 50 Mg $CaCO_3$, $(200 \times 0.000\,477) = 0.0954$ Mg NaOH and 200 Mg water. After concentration to 50% solids, mass of NaOH in solution

$$= (0.0954 \times 50/200.0954) = 0.0238 \text{ Mg}$$

and per cent $CaCO_3$ in dried solids

$$= (100 \times 50/50.0238) = \underline{\underline{99.95\%}}$$

This approach does not take into account the fact that filtrate is returned to the second thickener together with wash water. In this event, the calculation may be modified as follows:

The underflow product from the second thickener contains:

$$50 \text{ Mg } CaCO_3, \quad 200x_2 \text{ Mg NaOH} \quad \text{and} \quad 200 \text{ Mg water}$$

After filtration, the 50 Mg $CaCO_3$ is associated with 50 Mg solution of the same concentration and hence this contains:

$$50x_2/(1 + x_2) \text{ Mg NaOH} \quad \text{and} \quad 50/(1 + x_2) \text{ Mg water}$$

The remainder is returned with the overflow feed to the second thickener. The filtrate returned contains:

$$200x_2 - 50x_2/(1 + x_2)\, Mg\, NaOH$$

and

$$200 - 50/(1 + x_2)\, Mg\, water$$

The balances now become:

	CaCO₃	NaOH	Water
Overall			
Underflow feed	50	1	949
Overflow feed	—	$200x_2 - 50x_2/(1 + x_2)$	$400 - 50/(1 + x_2)$
Underflow product	50	$200x_2$	200
Overflow product	—	$1 - 50x_2/(1 + x_2)$	$1149 - 50/(1 + x_2)$
Thickener 1			
Underflow feed	50	1	949
Overflow feed	—	$200x_1 - 50x_2/(1 + x_2)$	$400 - 50/(1 + x_2)$
Underflow product	50	$200x_1$	200
Overflow product	—	$1 - 50x_2/(1 + x_2)$	$1149 - 50/(1 + x_2)$
Thickener 2			
Underflow feed	50	$200x_1$	200
Overflow feed	—	$200x_2 - 50x_2/(1 + x_2)$	$400 - 50/(1 + x_2)$
Underflow product	50	$200x_2$	200
Overflow product	—	$200x_1 - 50x_2/(1 + x_2)$	$400 - 50/(1 + x_2)$

As before, assuming equilibrium is attained,

$$x_2 = [200x_1 - 50x_2/(1 + x_2)]/[400 - 50/(1 + x_2)]$$
$$x_1 = [1 - 50x_2/(1 + x_2)]/[1149 - 50/(1 + x_2)]$$

Solving simultaneously,

$$x_1 = 0.000\,870\, Mg/Mg \quad \text{and} \quad x_2 = 0.000\,435\, Mg/Mg$$

The solid product leaving the filter contains 50 Mg CaCO₃

and $(50 \times 0.000\,435)/(1 + 0.000\,435) = 0.021\,75\, Mg\, NaOH$ in solution

After drying, the solid product will contain

$$(100 \times 50)/(50 + 0.021\,75) = \underline{99.96\%\, CaCO_3}$$

Problem 10.7

Barium carbonate is to be made by the interaction of sodium carbonate and barium sulphide. The quantities that are fed to the reaction agitators per 24 h day are as follows: 20 Mg of barium sulphide dissolved in 60 Mg of water together with the theoretically necessary amount of sodium carbonate.

There are three thickeners in series, run on a countercurrent decantation system. Overflow from the second thickener goes to the agitators and overflow from the first thickener is to be 10% sodium sulphide. Sludge from all thickeners carried two parts of water to one part of barium carbonate (w/w).

How much sodium sulphide will remain in the dried barium carbonate precipitate?

Solution

Basis: 1 day's operation

The reaction involved is:

$$BaS + Na_2CO_3 = BaCO_3 + Na_2S$$

$$169 \qquad 106 \qquad\quad 197 \qquad 78$$

Thus 20 Mg BaS will react to produce $(20 \times 197/169) = 23\cdot3$ Mg $BaCO_3$

and $\qquad\qquad\qquad\qquad\qquad (20 \times 78/169) = 9\cdot23$ Mg Na_2S

The calculation may be made on the basis of this material entering the washing thickeners together with 60 Mg water. Let x_1, x_2, and x_3 be the Na_2S/water ratio in the respective thickener. The mass balances are as follows:

	$BaCO_3$	Na_2S	Water
Overall			
Underflow feed	23·3	9·23	60
Overflow feed	—	—	w (say)
Underflow product	23·3	$46\cdot6x_3$	46·6
Overflow product	—	$9\cdot23 - 46\cdot6x_3$	$w + 13\cdot4$
Thickener 1			
Underflow feed	23·3	9·23	60
Overflow feed	—	$46\cdot6(x_1 - x_3)$	w
Underflow product	23·3	$46\cdot6x_1$	46·6
Overflow product	—	$9\cdot23 - 46\cdot6x_3$	$w + 13\cdot4$
Thickener 2			
Underflow feed	23·3	$46\cdot6x_1$	46·6
Overflow feed	—	$46\cdot6(x_2 - x_3)$	w
Underflow product	23·3	$46\cdot6x_2$	46·6
Overflow product	—	$46\cdot6(x_1 - x_3)$	w
Thickener 3			
Underflow feed	23·3	$46\cdot6x_2$	46·6
Overflow feed	—	—	w
Underflow product	23·3	$46\cdot6x_3$	46·6
Overflow product	—	$46\cdot6(x_2 - x_3)$	w

In the overflow product leaving the first thickener,

$$(9\cdot23 - 46\cdot6x_3)/(13\cdot4 + w + 9\cdot23 - 46\cdot6x_3) = 0\cdot10 \qquad\qquad \text{(i)}$$

Assuming equilibrium is attained in each thickener,

$$x_1 = (9\cdot23 - 46\cdot6x_3)/(13\cdot4 + w) \qquad\qquad \text{(ii)}$$

$$x_2 = 46\cdot6(x_1 - x_3)/w \qquad\qquad \text{(iii)}$$

$$x_3 = 46\cdot6(x_2 - x_3)/w \qquad\qquad \text{(iv)}$$

Solving equations (i)–(iv) simultaneously,

$$x_1 = 0\cdot112, \quad x_2 = 0\cdot066, \quad x_3 = 0\cdot030, \quad \text{and} \quad w = 57\cdot1 \text{ Mg/day}$$

In the underflow product from the third thickener, mass of Na_2S

$$= (46.6 \times 0.030) = 1.4\,Mg \text{ associated with } 23.3\,Mg\ BaCO_3$$

When this stream is dried, the barium carbonate will contain

$$(100 \times 1.4)/(1.4 + 23.3) = \underline{\underline{5.7\%}} \text{ sodium sulphide}$$

Problem 10.8

In the production of caustic soda by the action of calcium hydroxide on sodium carbonate, 1 kg/s of sodium carbonate is treated with the theoretical quantity of lime. The sodium carbonate is made up as a 20% solution. The material from the extractors is fed to a countercurrent washing system where it is treated with 2 kg/s of clean water. The washing thickeners are so arranged that the ratio of the volume of liquid discharged in the liquid offtake to that discharged with the solid is the same in all the thickeners and is equal to 4.0.

How many thickeners must be arranged in series so that not more than 1% of the sodium hydroxide discharged with the solid from the first thickener is wasted?

Solution

The reaction involved is:

$$Na_2CO_3 + Ca(OH)_2 = 2NaOH + CaCO_3$$
$$106 \qquad 74 \qquad 80 \qquad 100$$

Thus 1 kg/s Na_2CO_3 forms $(1 \times 80/106) = 0.755$ kg/s NaOH

and $\qquad (1 \times 100/106) = 0.943$ kg/s $CaCO_3$

In a 20% solution, 1 kg/s Na_2CO_3 is associated with $(1 - 0.20)/0.2 = 4.0$ kg/s water.

If x kg/s NaOH leaves in the underflow product from the first thickener, then $0.01x$ kg/s NaOH should leave in the underflow product from the nth thickener. The amount of NaOH in the overflow from the first thickener is then given by an overall balance as:

$$= 0.755 - 0.01x \text{ kg/s}$$

Since the volume of the overflow product is $4x$ (the volume of solution in underflow product),

$$0.755 - 0.01x = 4x$$

$$\therefore \qquad x = 0.188 \text{ kg/s}$$

and the NaOH leaving the nth thickener in the underflow $= (0.01 \times 0.188)$

$$= 0.001\,88 \text{ kg/s}$$

Thus the fraction of solute fed to the washing system which remains associated with the washed solids,

$$f = (0.001\,88/0.755) = 0.0025 \text{ kg/kg}$$

In this case $R = 4.0$ and in equation 10.17:

$$n = \{\log[1 + (4-1)/0.0025]\}/(\log 4) - 1$$

$$= 4.11, \quad \text{say } \underline{\underline{5 \text{ washing thickeners}}}$$

Problem 10.9

A plant produces 100 kg/s of titanium dioxide pigment which must be 99.9% pure when dried. The pigment is produced by precipitation and the material, as prepared, is contaminated with 1 kg of salt solution, containing 0.55 kg of salt, per kg of pigment. The material is washed countercurrently with water in a number of thickeners arranged in series. How many thickeners will be required if water is added at the rate of 200 kg/s and the solid discharged from each thickener removes 0.5 kg of solvent per kg of pigment?

What will be the required number of thickeners if the amount of solution removed in association with the pigment varies in the following way with the concentration of the solution in the thickener?

Concentration (solute/solution)	Solution/unit mass of pigment
0	0.30
0.1	0.32
0.2	0.34
0.3	0.36
0.4	0.38
0.5	0.40

The concentrated wash liquor is mixed with the material fed to the first thickener.

Solution

Part 1

Overall balance on plant in kg/s.

	TiO_2	Salt	Water
Feed from reactor	100	55	45
Wash liquor added	—	—	200
Washed solid	100	0.1	50
Liquid product	—	54.9	195

Solvent in underflow from final washing thickener $= 50$ kg/s.
Solvent in overflow will be the same as that supplied for washing, i.e. $= 200$ kg/s.

$$\frac{\text{Solvent discharged in overflow}}{\text{Solvent discharged in underflow}} = 4 \text{ for the washing thickeners.}$$

Liquid product from plant contains 54.9 kg of salt in 195 kg of solvent.
This ratio will be the same in the underflow from the first thickener.

Thus the material fed to the washing thickeners consists of 100 kg TiO_2, 50 kg solvent, and $(50 \times 54\cdot9/195) = 14$ kg salt.

Required number of thickeners for washing is given by equation 10.16:

$$\frac{4-1}{4^{n+1}-1} = \frac{0\cdot1}{14}$$

i.e.

$$4^{n+1} = 421$$

giving

$$4 < n+1 < 5$$

Thus 4 washing thickeners or a total of 5 thickeners are required.

Part 2

The same symbols will be used as in Volume 2.

By inspection of the data, it is seen that $W_{h+1} = 0\cdot30 + 0\cdot2X_h$.

Then

$$S_{h+1} = W_{h+1}X_h = 0\cdot30X_h + 0\cdot2X_h^2 = 5W_{h+1}^2 - 1\cdot5W_{h+1}$$

Consider the passage of unit quantity of TiO_2 through the plant.

$$L_{n+1} = 0, \qquad w_{n+1}' = 2, \qquad X_{n+1} = 0$$

since 200 kg/s pure solvent is used.

$$S_{n+1} = 0\cdot001 \quad \text{and therefore} \quad W_{n+1} = 0\cdot3007.$$

$$S_1 = 0\cdot55 \quad \text{and} \qquad W_1 = 1\cdot00$$

Thus the concentration in the first thickener is given by equation 10.23:

$$X_1 = \frac{L_{n+1} + S_1 - S_{n+1}}{w_{n+1} + W_1 - W_{n+1}} = \frac{0 + 0\cdot55 - 0\cdot001}{2 + 1 - 0\cdot3007} = \frac{0\cdot549}{2\cdot700} = 0\cdot203$$

From equation 10.26:

$$X_{h+1} = \frac{L_{n+1} - S_{n+1} + S_{h+1}}{w_{n+1} - W_{n+1} + W_{h+1}} = \frac{0 - 0\cdot001 + S_{h+1}}{2 - 0\cdot3007 + W_{h+1}} = \frac{-0\cdot001 + S_{h+1}}{1\cdot7 + W_{h+1}}$$

Since $X_1 = 0\cdot203$, $\qquad W_2 = 0\cdot30 + 0\cdot2 \times 0\cdot203 = 0\cdot3406$

and $\qquad\qquad\qquad S_2 = 0\cdot3406 \times 0\cdot203 = 0\cdot0691$

Thus $\qquad\qquad X_2 = \dfrac{0\cdot0691 - 0\cdot001}{1\cdot7 + 0\cdot3406} = \dfrac{0\cdot0681}{2\cdot0406} = 0\cdot0334$

Since $X_2 = 0\cdot0334$, $\qquad W_3 = 0\cdot30 + 0\cdot2 \times 0\cdot0334 = 0\cdot30668$

and $\qquad\qquad\qquad S_2 = 0\cdot3067 \times 0\cdot0334 = 0\cdot010\,25$

Thus $\qquad\qquad X_3 = \dfrac{0\cdot010\,25 - 0\cdot001}{1\cdot7 + 0\cdot3067} = \dfrac{0\cdot009\,25}{2\cdot067} = 0\cdot004\,47$

Since $X_3 = 0\cdot004\,47$, $\qquad W_4 = 0\cdot300\,89 \quad \text{and} \quad S_4 = 0\cdot0013$

Hence, by the same method, $\qquad X_4 = 0\cdot000\,150$

Since $X_4 = 0.000\,150$, $W_5 = 0.300\,03$ and $S_5 = 0.000\,045$

Thus S_5 is less than S_{n+1} and therefore 4 thickeners are required.

Problem 10.10

Prepared cottonseed meats containing 35% extractable oil are fed to a continuous countercurrent extractor of the intermittent drainage type using hexane as solvent. The extractor consists of ten sections, the section efficiency being 50%. The entrainment, assumed constant, is 1 kg solution/kg solids.

What will be the oil concentration in the outflowing solvent if the extractable oil content in the meats is to be reduced to 0.5% by weight?

Solution

Basis: 100 kg inert cottonseed material

Mass of oil in underflow feed = $(100 \times 0.35)/(1 - 0.35) = 53.8$ kg.

In the underflow product from the plant, mass of inerts = 100 kg and hence mass of oil = $(100 \times 0.005)/(1 - 0.005) = 0.503$ kg.

This is within 100 kg solution and hence mass of hexane in underflow product = $(100 - 0.503) = 99.497$ kg.

An *overall balance* is now:

	Inerts	Oil	Hexane
Underflow feed	100	53.8	—
Overflow feed	—	—	h (say)
Underflow product	100	0.503	99.497
Overflow product	—	53.297	$(h - 99.497)$

Since there are ten stages, each 50% efficient, the system may be treated as consisting of five theoretical stages each of 100% efficiency, i.e. equilibrium is attained in each stage. On this basis, the underflow from stage 1 contains 100 kg solution in which the oil/hexane ratio = $53.297/(h - 99.497)$ and hence the amount of oil in this stream,

$$S_1 = 100[1 - (h - 99.497)/(h - 46.2)] \text{ kg}$$

$$S_{n+1} = 0.503 \text{ kg}$$

With constant underflow, the amount of solution in the overflow from each stage = h kg and the solution in the underflow = 100 kg.

$$\therefore \qquad R = h/100 = 0.01h$$

and in equation 10.16:

$$0.503/[100 - 100(h - 99.497)/(h - 46.2)] = (0.01h - 1)/[(0.01h)^5 - 1]$$

$$\therefore \qquad (0.503h - 23.24) = (53.30h - 5330)/[(0.01h)^5 - 1]$$

Solving by trial and error, $h = 238\,\text{kg}$

∴ In the overflow product,

$$\text{mass of hexane} = (238 - 99\cdot497) = 138\cdot503\,\text{kg}$$

$$\text{mass of oil} = 53\cdot298\,\text{kg}$$

and $\qquad \text{concentration of oil} = (100 \times 53\cdot297)/(53\cdot297 + 138\cdot503)$

$$= \underline{\underline{27\cdot8\%}}$$

Problem 10.11

Seeds containing 25% by weight of oil are extracted in a countercurrent plant, and 90% of the oil is to be recovered in a solution containing 50% of oil. It has been found experimentally that the amount of solution removed in the underflow in association with every kilogram of insoluble matter is given by the equation:

$$k = 0\cdot7 + 0\cdot5y_s + 3y_s^2$$

where y_s is the concentration of the overflow solution (weight fraction of solute).
    If the seeds are extracted with fresh solvent, how many ideal stages are required?

Solution

Basis: 100 kg underflow feed to the first stage

The initial step is to obtain the underflow line, i.e. a plot of x_s against x_A and the calculations are made as follows:

y_s	k	Ratio (kg/kg inerts)			Mass fraction	
		Oil (ky_s)	Solvent $k(1-y_s)$	Underflow $(k+1)$	Oil (x_A)	Solvent (x_s)
0	0·70	0	0·70	1·70	0	0·412
0·1	0·78	0·078	0·702	1·78	0·044	0·394
0·2	0·92	0·184	0·736	1·92	0·096	0·383
0·3	1·12	0·336	0·784	2·12	0·159	0·370
0·4	1·38	0·552	0·828	2·38	0·232	0·348
0·5	1·70	0·850	0·850	2·70	0·315	0·315
0·6	2·08	1·248	0·832	3·08	0·405	0·270
0·7	2·52	1·764	0·756	3·52	0·501	0·215
0·8	3·02	2·416	0·604	4·02	0·601	0·150
0·9	3·58	3·222	0·358	4·58	0·704	0·078
1·0	4·20	4·20	0	5·20	0·808	0

The curve of x_A and x_s is drawn in Fig. 10b.
    Considering the *underflow feed*, the seeds contain 25% oil, 75% inerts, and the point $x_{s1} = 0$, $x_{A1} = 0\cdot25$ is drawn in as x_1.
    In the *overflow feed*, pure solvent is used and hence

$$y_{s \cdot n+1} = 1\cdot0, \qquad y_{A \cdot n+1} = 0$$

This point is marked as y_{n+1}.

In the *overflow product*, the oil concentration is 50% and

$$y_{s1} = 0.50, \qquad y_{A1} = 0.50$$

This point lies on the hypotenuse and is marked y_1.

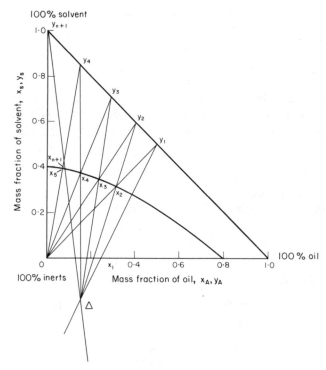

FIG. 10b

90% of the oil is recovered, leaving $25(1-0.90) = 2.5$ kg in the underflow product associated with 75 kg inerts,

i.e.   ratio (oil/inerts) $= 2.5/75 = 0.033 = ky_s$

∴ $0.033 = 0.7y_s + 0.5y_s^2 + 3y_s^3$

Solving by substitution,

$$y_s = 0.041 \quad \text{and hence} \quad k = (0.033/0.041) = 0.805$$

∴ $x_A = 0.0173 \quad \text{and} \quad x_s = 0.405$

This point is drawn as x_{n+1} on the x_s vs. x_A curve.

The pole point Δ is obtained where $y_{n+1} \cdot x_{n+1}$ and $y_1 \cdot x_1$ extended meet and the construction described in Chapter 10 is then followed.

It is found that x_{n+1} lies between x_4 and x_5 and hence

4 ideal stages are required

Problem 10.12

Halibut oil is extracted from granulated halibut livers in a countercurrent multi-batch arrangement using ether as solvent. The solids charge contains 0·35 kg oil/kg exhausted livers and it is desired to obtain a 90% oil recovery. How many theoretical stages are required if 50 kg ether is used per 100 kg untreated solids? The entrainment data are as follows:

Concentration of overflow (kg oil/kg solution)	0	0·1	0·2	0·3	0·4	0·5	0·6	0·67
Entrainment (kg solution/kg extracted livers)	0·28	0·34	0·40	0·47	0·55	0·66	0·80	0·96

Solution

The entrainment data may be converted into terms of mass fractions as follows:

Overflow concentration (kg oil/kg soln.)	Entrainment (kg soln./ kg livers)	Ratio (kg/kg extracted livers) Oil	Ether	Underflow	Mass fraction x_A	x_S
0	0·28	0	0·280	1·280	0	0·219
0·1	0·34	0·034	0·306	1·340	0·025	0·228
0·2	0·40	0·080	0·320	1·400	0·057	0·228
0·3	0·47	0·141	0·329	1·470	0·096	0·223
0·4	0·55	0·220	0·330	1·550	0·142	0·212
0·5	0·66	0·330	0·330	1·660	0·199	0·198
0·6	0·80	0·480	0·320	1·880	0·255	0·170
0·67	0·96	0·643	0·317	1·960	0·328	0·162

and these are plotted in Fig. 10c.

On the *basis of 100 kg untreated solids*:

In the *underflow feed*, 0·35 kg oil is associated with each kg exhausted livers.

∴ mass of livers fed $= 100/(1+0·35) = 74$ kg containing $(100-74) = 26$ kg oil

∴ $$x_A = 0·26, \qquad x_s = 0$$

This point is marked in as x_1.

In the *overflow feed*, pure ether is used and $y_s = 1·0$, $x_s = 0$, which is marked in as y_{n+1}.

Since the recovery of oil is 90%, the overall mass balance becomes:

	Exhausted livers	Oil	Ether
Underflow feed	74	26	—
Overflow feed	—	—	50
Underflow product	74	2·6	e (say)
Overflow product	—	23·4	$(50-e)$

In the *underflow product*,

the ratio (oil/exhausted livers) $= (2\cdot6/74) = 0\cdot035\,\text{kg/kg}$

which from the entrainment data is equivalent to

$$x_A = 0\cdot025, \quad x_s = 0\cdot228$$

which is marked in as x_{n+1}.

The ratio (ether/exhausted livers) $= 0\cdot306\,\text{kg/kg}$ or $e = (0\cdot306 \times 74) = 22\cdot6\,\text{kg}$

In the *overflow product*,

the mass of ether $= (50 - 22\cdot6) = 27\cdot4\,\text{kg}$

and $$y_A = 23\cdot4/(23\cdot4 + 27\cdot4) = 0\cdot46, \quad y_S = 0\cdot54$$

which is marked in as y_1.

Following the construction described in Chapter 10, Section 10.6.3, it is found that point x_4 coincides exactly with x_{n+1} as shown in Fig. 10c and hence

<u>3 ideal stages are required</u>

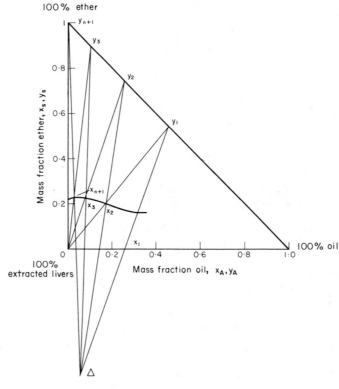

Fig. 10c

SECTION 11

DISTILLATION

Problem 11.1

A liquor of four components A, B, C, and D with 0·3 mol fraction each of A, B, and C is to be continuously fractionated to give a top product of 0·9 mol fraction A and 0·1 mol fraction B. The bottoms are to contain not more than 0·05 mol fraction A. Estimate the minimum reflux ratio required for this separation if the relative volatility of A to B is 2·0.

Solution

The given data may be tabulated as shown below:

	Feed	Top	Bottoms
A	0·3	0·9	0·05
B	0·3	0·1	
C	0·3	—	
D	0·1	—	

The equations of Underwood and Fenske may be used to find the minimum number of plates and the minimum reflux ratio for a binary system. For a multicomponent system n_m may be found by using the two key components in place of the binary system and the relative volatility between those components in equation 11.40 enables the minimum reflux ratio R_m to be found. Using the feed and top compositions of component A:

$$R_m = \frac{1}{\alpha - 1}\left[\left(\frac{x_d}{x_f}\right) - \alpha\frac{(1 - x_d)}{(1 - x_f)}\right] \qquad \text{(equation 11.40)}$$

$$\therefore \quad R_m = \frac{1}{2 - 1}\left[\left(\frac{0\cdot9}{0\cdot3}\right) - 2\frac{(1 - 0\cdot9)}{(1 - 0\cdot3)}\right]$$

$$= \underline{\underline{2\cdot71}}$$

Problem 11.2

During the batch distillation of a binary mixture in a packed column the product had 0·60 mol fraction of the more volatile component when the concentration in the still was 0·40 mol fraction. If the reflux ratio in use was 20:1 and the vapour composition y

is related to the liquor composition x by the equation $y = 1 \cdot 035x$ over the range of concentration concerned, determine the number of ideal plates represented by the column (x and y are in mol fractions).

Solution

It is seen in equation 11.32, the equation of the operating line, that the slope is given by $R/(R+1)(=L/V)$ and the intercept on the y-axis by

$$x_d/(R+1) = (D/V_n)$$

$$y_n = \frac{R}{R+1}x_{n+1} + \frac{x_D}{R+1}$$ (equation 11.32)

In this problem, the equilibrium curve over the range $x = 0 \cdot 40$ to $x = 0 \cdot 60$ is given by $y = 1 \cdot 035x$ so that it may be drawn as shown in Fig. 11a. The intercept of the operating line on the y-axis is equal to

$$x_d/(R+1) = 0 \cdot 60/(20+1) = 0 \cdot 029$$

so that the operating line is drawn through the points $(0 \cdot 60, 0 \cdot 60)$ and $(0, 0 \cdot 029)$ as shown.

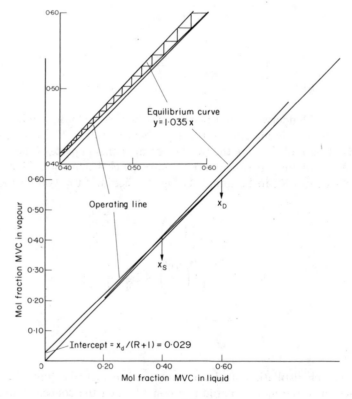

FIG. 11a

In this particular example, all these lines are closely spaced so that the relevant section is enlarged in the inset of Fig. 11a. By stepping off the theoretical plates by the McCabe–Thiele method, it is seen that 18 theoretical plates are represented by the column.

Problem 11.3

A mixture of water and ethyl alcohol containing 0·16 mol fraction alcohol is continuously distilled in a plate fractionating column to give a product containing 0·77 mol fraction alcohol and a waste of 0·02 mol fraction alcohol. It is proposed to withdraw 25% of the alcohol in the entering stream as a side stream with a mol fraction of 0·50 alcohol.

Determine the number of theoretical plates required and the plate from which the side stream should be withdrawn if the feed is liquor at the boiling-point and a reflux ratio of 2 is used.

Solution

Taking a basis of 100 kmol of feed to the column, 16 kmol of alcohol enter, and 25%, i.e. 4 kmol, are to be removed in the side stream. As the side-stream composition is to be 0·5, there are 8 kmol in that stream.

An overall mass balance gives: $F = D + W + S$

i.e. $$100 = D + W + 8$$

or $$92 = D + W$$

A mass balance on the alcohol gives:

$$100 \times 0·16 = 0·77D + 0·02W + 4$$

or $$12 = 0·77D + 0·02W$$

from which $D = $ kmol distillate $= 13·55$

and $W = $ kmol bottoms $= 78·45$

In the top section between the side stream and the top of the column:

$$R = L_n/D = 2, \quad \text{hence } L_n = 27·10$$

$$V_n = L_n + D, \quad \text{giving } V_n = 40·65$$

For the section between the feed and the side stream:

$$V_s = V_n = 40·65$$

$$L_n = S + L_s$$

$\therefore$ $$L_s = 27·10 - 8 = 19·10$$

In the bottom of the column:

$$L_m = L_s + F = 19·10 + 100 = 119·10$$

if the feed is at its boiling-point.

Chemical Engineering

$$V_m = L_m - W = 119 \cdot 10 - 78 \cdot 45 = 40 \cdot 65$$

The slope of the operating line is always L/V so that the slope in each part of the column can now be calculated. The top operating line passes through the point (x_d, x_d) and has a slope of $27 \cdot 10/40 \cdot 65 = 0 \cdot 67$. This is shown in Fig. 11b and applies until $x_s = 0 \cdot 50$ where the slope becomes $19 \cdot 10/40 \cdot 65 = 0 \cdot 47$. The operating line in the bottom of the column applies from $x_f = 0 \cdot 16$ and passes through the point (x_w, x_w) with a slope $119 \cdot 10/40 \cdot 65 = 2 \cdot 92$.

The steps corresponding to the theoretical plates may be drawn as shown when 8 plates are required with the side stream withdrawn from the fourth plate from the top.

Problem 11.4

In a mixture to be fed to a continuous distillation column, the mol fraction of phenol is 0·35, of *o*-cresol 0·15, of *m*-cresol 0·30, and of xylenes 0·20. It is hoped to obtain a product with a mol fraction of phenol 0·952, of *o*-cresol 0·0474, and of *m*-cresol 0·0006. If the volatility relative to *o*-cresol of phenol is 1·26 and of *m*-cresol 0·70, estimate how many theoretical plates would be required at total reflux.

Solution

The data given are tabulated below:

Component	Feed	Top	Bottoms	α
P	0·35	0·952		1·26
O	0·15	0·0474		1·0
M	0·30	0·0006		0·7
X	0·20	—		
		1·0000		

Fenske's equation may be used to find the minimum number of plates, i.e. the number of plates at total reflux

$$n + 1 = \frac{\log \left[(x_A/x_B)_d (x_B/x_A)_s \right]}{\log \alpha_{AB}} \qquad \text{(equation 11.42)}$$

For multicomponent systems, components A and B refer to the light and heavy keys respectively. In this problem *o*-cresol is the light key and *m*-cresol is the heavy key. A mass balance must be performed to determine the bottom composition. Taking a basis of 100 kmol of feed:

$$100 = D + W$$

For phenol, $\qquad 100 \times 0 \cdot 35 = 0 \cdot 952D + x_{wp} W$

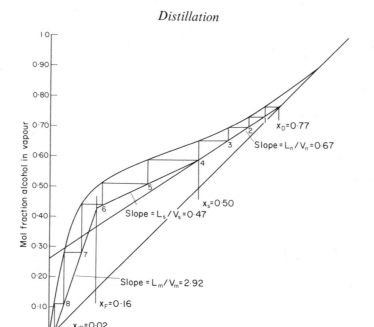

Fig. 11b

If x_{wp} is taken as zero (a reasonable assumption) then:

$$D = 36\cdot8 \qquad W = 63\cdot2$$

For *o*-cresol: $100 \times 0\cdot15 = 0\cdot0474 \times 36\cdot8 + x_{wo} \times 63\cdot2$

and $$x_{wo} = 0\cdot21$$

For *m*-cresol: $100 \times 0\cdot30 = 0\cdot0006 \times 36\cdot8 + x_{wm} \times 63\cdot2$

and $$x_{wm} = 0\cdot474$$

By difference: $$x_{wx} = 0\cdot316$$

$$\alpha_{om} = 1/0\cdot7 = 1\cdot43$$

Hence substituting into Fenske's equation gives:

$$n + 1 = \frac{\log\left[(0\cdot0474/0\cdot0006)(0\cdot474/0\cdot21)\right]}{\log 1\cdot43}$$

and $$n = 13\cdot5$$

Problem 11.5

A continuous fractionating column, operating at atmospheric pressure, is to be designed to separate a mixture containing 15·67% CS_2 and 84·33% CCl_4 into an over-head product containing 91% CS_2 and a waste of 97·3% CCl_4 (all weight per cent).

Assume a plate efficiency of 70% and a reflux of 3·16 kmol per kmol of product.
Using the data below, determine the number of plates required.
Feed enters at 290 K with a specific heat of 1·7 kJ/kg K and boiling-point of 336 K.
Latent heat of CS_2 and CCl_4 is 25,900 kJ/kmol.

Mol per cent CS_2 in vapour:

   0   8·23   15·55   26·6   33·2   49·5   63·4   74·7   82·9   87·8   93·2

Mol per cent CS_2 in liquor:

   0   2·96   6·15   11·06   14·35   25·85   39·0   53·18   66·30   75·75   86·04

Solution

The equilibrium data are presented in Fig. 11c and the problem will be solved using
the method of McCabe and Thiele. All compositions must be in terms of mol fractions
so that:

Top product
$$x_d = \frac{(91/76)}{(91/76) + (9/154)} = 0.953$$

Feed
$$x_f = \frac{(15.67/76)}{(15.67/76) + (84.33/154)} = 0.274$$

Bottom product
$$x_w = \frac{(2.7/76)}{(2.7/76) + (97.3/154)} = 0.053$$

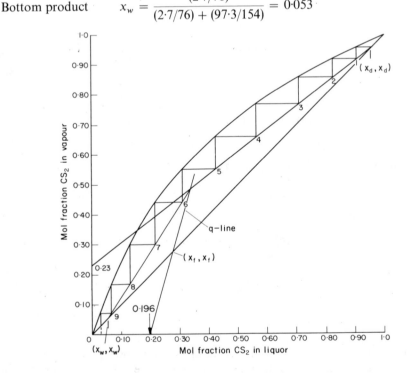

Fɪɢ. 11c

In this problem, the feed is not at its boiling-point so the slope of the q-line must be determined in order to locate the intersection of the operating lines.

q is defined as the heat to vaporise 1 kmol of feed/molar latent heat of feed:

$$q = (\lambda + H_{fs} - H_f)/\lambda$$

where λ is the molar latent heat, H_{fs} is the enthalpy of 1 kmol of feed at its boiling-point, and H_f is the enthalpy of 1 kmol of feed.

The feed composition is 27·4% CS_2 and 72·6% CCl_4 so that the mean molecular weight of the feed

$$= (0·274 \times 76) + (0·726 \times 154)$$

$$= 132·6 \text{ kg/kmol}$$

Taking a datum of 273 K:

$$H_f = 1·7 \times 132·6(290 - 273) = 3832 \text{ kJ/kmol}$$

$$H_{fs} = 1·7 \times 132·6(336 - 273) = 14{,}200 \text{ kJ/kmol}$$

$$\lambda = 25{,}900 \text{ kJ/kmol}$$

$$\therefore \quad q = (25{,}900 + 14{,}200 - 3832)/25{,}900$$

$$= 1·4$$

The intercept of the q-line on the x-axis is shown from equation 11.30 to be x_f/q:

$$y_q = \frac{q}{q-1} x_q - \frac{x_f}{q-1} \qquad \text{(equation 11.30)}$$

$$x_f/q = 0·274/1·4 = 0·196$$

Thus the q-line is drawn through (x_f, x_f) and $(0·196, 0)$ as shown in Fig. 11c. As the reflux ratio is given as 3·16, the top operating line may be drawn through (x_d, x_d) and $(0, x_d/4·16)$. The lower operating line is drawn by joining the intersection of the top operating line and the q-line with the point (x_w, x_w).

The theoretical plates may be stepped off as shown and 9 theoretical plates are shown in Fig. 11c. If the plate efficiency is 70%, the number of actual plates $= 9/0·7 = 12·85$,

i.e. <u>13 plates are required</u>

Problem 11.6

A batch fractionation is carried out in a small column which has the separating power of 6 theoretical plates. The mixture consists of benzene and toluene with 0·60 mol fraction of benzene. A distillate is required, of constant composition, of 0·98 mol fraction benzene and the operation is discontinued when 83% of the benzene charged has been removed as distillate. Estimate the reflux ratio needed at the start and finish of the distillation if the relative volatility of benzene to toluene is taken as 2·46.

Solution

The equilibrium data are calculated from the relative volatility from:

$$y_A = \frac{\alpha x_A}{1 + (\alpha - 1) x_A}$$

(equation 11.8)

to give:

x_A	0	0·1	0·2	0·3	0·4	0·5	0·6	0·7	0·8	0·9	1·0
y_A	0	0·215	0·380	0·513	0·621	0·711	0·787	0·852	0·908	0·956	1·0

If a constant product is to be obtained from a batch still, the reflux ratio must be constantly increased. Initially S_1 kmol of liquor are in the still with a composition x_{s1} of the MVC and a reflux ratio of R_1 is required to give the desired product composition x_d. Later, when S_2 kmol remain in the still of composition x_{s2}, the reflux ratio has increased to R_2 when the amount of product is D kmol.

From an overall mass balance:

$$S_1 - S_2 = D$$

For the MVC:

$$S_1 x_{s1} - S_2 x_{s2} = D x_d$$

from which

$$D = S_1 \frac{(x_{s1} - x_{s2})}{(x_d - x_{s2})}$$

(equation 11.75)

In this problem $x_{s1} = 0·6$ and $x_d = 0·98$ and there are 6 theoretical plates in the column. It remains, by using the equilibrium data, to determine values of x_{s2} for selected reflux ratios. This is done graphically by choosing an intercept on the y-axis, calculating R, drawing the resulting operating line, and stepping off in the normal way 6 theoretical plates and finding the still composition x_{s2}.

This is shown in Fig. 11d for two very different reflux ratios and the procedure is repeated to enable the following table to be produced.

Intercept on y-axis (ϕ)	Reflux ratio ($\phi = x_d/R + 1$)	x_{s2}
0·45	1·18	0·725
0·40	1·20	0·665
0·30	2·27	0·545
0·20	3·90	0·46
0·10	8·8	0·31
0·05	18·6	0·26

From the inset plot of x_{s2} vs. R:
At the start: $x_{s2} = 0·6$ and $R = 1·7$.

At the end: x_{s2} is calculated using equation 11.75 as follows:
If $S_1 = 100$ kmol
  kmol of benzene initially $= 100 \times 0·60 = 60$ kmol.
  kmol of benzene removed $= 0·83 \times 60 = 49·8$ kmol.

$$D = 49\cdot8/0\cdot98 = 50\cdot8$$

and

$$50\cdot8 = 100\frac{(0\cdot6 - x_{s2})}{(0\cdot98 - x_{s2})}$$

from which

$$x_{s2} = 0\cdot207 \quad \text{and} \quad \underline{\underline{R = 32}}$$

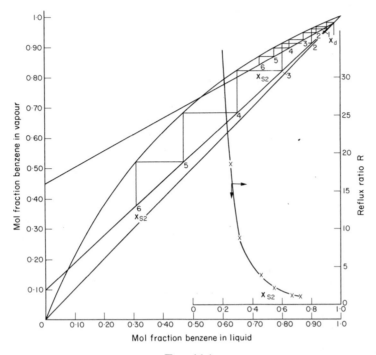

FIG. 11d

Problem 11.7

A continuous fractionating column is required to separate a mixture containing 0·695 mol fraction n-heptane (C_7H_{16}) and 0·305 mol fraction n-octane (C_8H_{18}) into products of 99 mol% purity. The column is to operate at a pressure of 101·3 kN/m² with a vapour velocity of 0·6 m/s. The feed is all liquid at its boiling-point, and is supplied to the column at 1·25 kg/s. The boiling-point at the top of the column may be taken as 372 K and the equilibrium data are:

y = mol fraction heptane in vapour:

| 0·96 | 0·91 | 0·83 | 0·74 | 0·65 | 0·50 | 0·37 | 0·24 |

x = mol fraction heptane in liquid:

| 0·92 | 0·82 | 0·69 | 0·57 | 0·46 | 0·32 | 0·22 | 0·13 |

Determine the minimum reflux ratio that will be required. What diameter column would be required if the reflux used were twice the minimum possible?

Solution

The equilibrium curve is plotted in Fig. 11e. As the feed is at its boiling-point, the *q*-line is vertical and the minimum reflux ratio may be found by joining the point (x_d, x_d) with the intersection of the *q*-line and the equilibrium curve. This line when produced to the *y*-axis gives an intercept of 0·475.

$$\therefore \qquad\qquad 0\cdot475 = x_D/(R_m + 1)$$

and
$$R_m = \underline{\underline{1\cdot08}}$$

If $2R_m$ is used,
$$R = 2\cdot16 \quad \text{and} \quad L_n/D = 2\cdot16$$

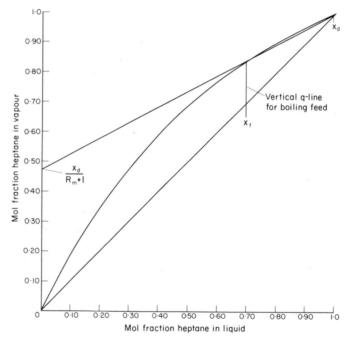

FIG. 11e

Taking a basis of 100 kmol of feed, mass balances overall and on the n-heptane give:

$$100 = D + W$$

and
$$100 \times 0\cdot695 = 0\cdot99D + 0\cdot01W$$

since 99% n-octane is required.

Hence
$$D = 69\cdot9 \quad \text{and} \quad W = 30\cdot1$$

$$\therefore \qquad L_n = 2\cdot16D = 151 \quad \text{and} \quad V_n = L_n + D = 221$$

The mean molecular weight of the feed $= (0\cdot695 \times 90) + (0\cdot305 \times 114)$

$$= 97\cdot3 \text{ kg/kmol}$$

$$\therefore \qquad \text{feed rate} = 1\cdot25/97\cdot3 = 0\cdot0128\,\text{kmol/s}$$

The vapour flow at the top of the column $= (221/100) \times 0\cdot0128$

$$= 0\cdot028\,\text{kmol/s}$$

The vapour density at the top of the column $= (1/22\cdot4)(273/372)$

$$= 0\cdot0328\,\text{kmol/m}^3$$

Hence the volumetric vapour flow $= 0\cdot028/0\cdot0328$

$$= 0\cdot853\,\text{m}^3/\text{s}$$

If vapour velocity $= 0\cdot6\,\text{m/s}$, the area required $= 0\cdot853/0\cdot6$

$$= 1\cdot42\,\text{m}^2$$

giving a column diameter of 1·34 m.

Problem 11.8

The vapour pressures of chlorobenzene and water are as follows:

Vapour pressure (kN/m²)	13·3	6·7	4·0	2·7
(mmHg)	100	50	30	20
Temperatures (K)				
Chlorobenzene	343·6	326·9	315·9	307·7
Water	324·9	311·7	303·1	295·7

A still is operated at a pressure of $18\,\text{kN/m}^2$, steam being blown continuously into it. Estimate the temperature of the boiling liquid and the composition of the distillate if liquid water is present in the still.

Solution

For steam distillation, assuming the gas laws to apply, the composition of the vapour produced can be obtained from the following:

$$\frac{m_A}{M_A}\bigg/\frac{m_B}{M_B} = \frac{P_A}{P_B} = \frac{y_A}{y_B} = \frac{P_A}{P - P_A} \qquad \text{(equation 11.97)}$$

where the subscript A refers to the component being recovered and B to steam, and m is the mass; M is the molecular weight; P_A, P_B is the partial pressure of A and B; and P is the total pressure.

If there is no liquid phase present, then from the phase rule there will be two degrees of freedom, and both the total pressure and the operating temperature can be fixed independently, and $P_B = P - P_A$ (which must not exceed the vapour pressure of pure water if no liquid phase is to appear).

With a liquid water phase present, there will only be one degree of freedom, and setting the temperature or pressure fixes the system, the water and the other component each exerting a partial pressure equal to its vapour pressure at the boiling-point of the mixture. In this case, the distillation temperature will always be less than that of boiling water at the total pressure in question. Consequently, a high-boiling organic material can be steam-distilled at temperatures below 373 K at atmospheric pressure. By using reduced operating pressures, the distillation temperature may be reduced still further, with consequent economy of steam.

A convenient method of calculating the temperature and composition of the vapour, for the case where the liquid water phase is present, is by use of Fig. 11.45 in Vol. 2 where the parameter $(P - P_B)$ is plotted for total pressures of 760, 300, and 70 mmHg, and the vapour pressures of a number of other materials are plotted directly against temperature. The intersection of the two appropriate curves gives the temperature of distillation and the molar ratio of water to organic material will be given by $(P - P_A)/P_A$.

The relevance of this method to this problem is illustrated in Fig. 11f where the vapour pressure of chlorobenzene is plotted as a function of temperature. On the same graph $P - P_B$ is plotted where $P = 18 \text{ kN/m}^2$ (130 mmHg) and P_B is the vapour pressure of water at the particular temperature. These curves are seen to intersect at the distillation temperature of 323 K.

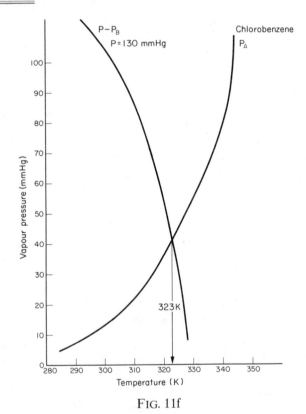

FIG. 11f

The composition of the distillate is found by substitution in equation 11.97 since $P_A = 41 \text{ mmHg}$ at 323 K. Hence

$$\frac{y_A}{y_B} = \frac{P_A}{P - P_A} = \frac{41}{130 - 41} = 0.461 \quad \text{(in mol fractions)}$$

Problem 11.9

The following figures represent the equilibrium conditions in mol fraction of benzene in benzene–toluene mixtures at their boiling-point.

Liquid	Vapour
0·51	0·72
0·38	0·60
0·26	0·45
0·15	0·30

If the liquid compositions on four adjacent plates in a column were 0·18, 0·28, 0·41, and 0·57 under conditions of total reflux, determine the plate efficiencies.

Solution

The equilibrium data are plotted in Fig. 11g over the range given and a graphical representation of the plate efficiency is shown in the inset. The efficiency E_{MI} in terms of the liquid compositions is defined by:

$$E_{MI} = \frac{x_{n+1} - x_n}{x_{n+1} - x_e} \qquad \text{(equation 11.127)}$$

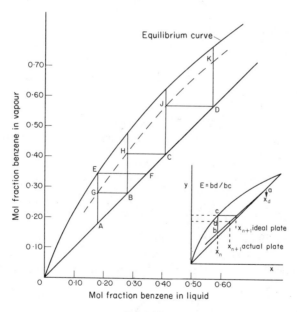

FIG. 11g

On the inset, the line *ab* represents an operating line and *bc* is the enrichment achieved on a theoretical plate. *bd* is the enrichment achieved on an actual plate so that the efficiency is then the ratio *bd/bc*.

Referring to the data given, at total reflux the conditions on actual plates in the column are shown as points A, B, C, and D. Considering point A, if equilibrium were achieved on that plate, point E would represent the vapour composition and point F the liquid composition on the next plate. However, the liquid on the next plate is determined by B so that the line AGE may be located and the efficiency is given by $AG/AE = 0.59$ or 59%.

In an exactly similar way, points H, J, and K are located to give efficiencies = 66%, 74%, and 77%.

Problem 11.10

A continuous rectifying column treats a mixture consisting of 40% of benzene by weight and 60% of toluene at the rate of 4 kg/s, and separates it into a product containing 97% of benzene and a liquid containing 98% toluene. The feed is liquid at its boiling-point.

(a) Calculate the weights of distillate and waste liquor per unit time.
(b) If a reflux ratio of 3.5 to 1 is employed, how many plates are required in the rectifying part of the column?
(c) What is the actual number of plates if the plate efficiency is 60%?

Mol fraction of benzene in liquid:

0·1	0·2	0·3	0·4	0·5	0·6	0·7	0·8	0·9

Mol fraction of benzene in vapour:

0·22	0·38	0·51	0·63	0·7	0·78	0·85	0·91	0·96

Solution

The equilibrium data are plotted on Fig. 11h. As the compositions are given as weight %, they must first be converted to mol fractions before the McCabe–Thiele method can be used.

$$\text{Mol fraction benzene in feed} = \frac{(40/78)}{(40/78) + (60/92)} = 0.440 = x_f$$

Similarly, $x_d = 0.974$ and $x_w = 0.024$

As the feed is a liquid at its boiling-point, the q-line is vertical and may be drawn at $x_f = 0.44$.

(a) A mass balance over the column and on the more volatile component gives in terms of mass flow rates:

$$4.0 = W' + D'$$

$$4 \times 0.4 = 0.02W' + 0.97D'$$

from which W' gives = bottoms flow rate = 2.4 kg/s

and D' = top product rate = 1.6 kg/s

(b) If $R = 3.5$, the intercept of the top operating line on the y-axis is given by $x_d/(R+1) = 0.974/4.5 = 0.216$, and thus the operating lines may be drawn as shown in Fig. 11h. The plates are stepped off as indicated and 10 theoretical plates are required.

(c) If the efficiency is 60%, the number of actual plates

$$= 10/0.6 = 16.7$$

or

17 actual plates

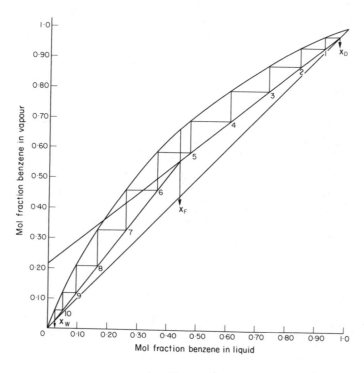

FIG. 11h

Problem 11.11

A distillation column is fed with a mixture of benzene and toluene in which the mol fraction of benzene is 0.35. The column is to yield a product in which the mol fraction of benzene is 0.95, when working with a reflux ratio of 3.2 to 1.0, and the waste from the column is not to exceed 0.05 mol fraction of benzene.

Assuming the plates used have an efficiency of 60%, find the number of plates required and the position of the feed point.

The relation between the mol fraction of benzene in liquid and in vapour is given by the data overleaf.

Mol fraction of benzene in liquid:

0·1	0·2	0·3	0·4	0·5	0·6	0·7	0·8	0·9

Mol fraction of benzene in vapour:

0·20	0·38	0·51	0·63	0·71	0·78	0·85	0·91	0·96

Solution

The solution to this problem is very similar to that of 11.10 except that the data are presented here in terms of mol fraction. Following a similar approach, the theoretical plates are stepped off and it is seen from Fig. 11i that 10 are required. Thus $10/0·6 = 16·7$ actual plates are required so that <u>17</u> would be employed and the feed tray corresponds to between ideal trays 5 and 6, so in practice the <u>eighth</u> actual tray from the top would be employed.

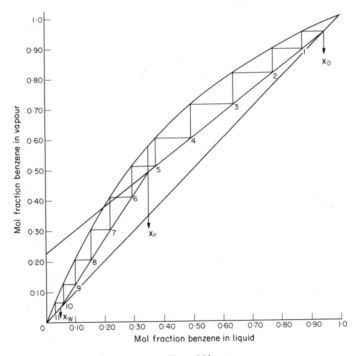

Fig. 11i

Problem 11.12

The accompanying table gives the relationship between the mol fraction of carbon disulphide in the liquid and in the vapour evolved from the mixture during the distillation of a carbon disulphide–carbon tetrachloride mixture:

x	0	0·20	0·40	0·60	0·80	1·00
y	0	0·445	0·65	0·795	0·91	1·00

Determine graphically the theoretical number of plates required for the rectifying and stripping portions of the column, using the following data:

Reflux ratio, 3 to 1.
Slope of fractionating line, 1·4.
Purity of product, 99%.
Percentage carbon disulphide in waste liquors, 1%.

What is the minimum slope of the rectifying line in this case?

Solution

The equilibrium data are plotted in Fig. 11j. In this problem no data have been provided on the composition or the nature of the feed so that conventional location of the q-line is impossible. The rectifying line may be drawn, however, as both the top composition and the reflux ratio are known. The intercept on the y-axis is given by $x_d/(R+1) = 0·99/4 = 0·248$.

The slope of the lower operating line is given as 1·4 so that the line may be drawn through the point (x_w, x_w) and the number of theoretical plates may be determined as shown to be 12.

The minimum slope of the rectifying line corresponds to an infinite number of theoretical stages. If the slope of the stripping line remains constant, then production of that line to the equilibrium curve enables the rectifying line to be drawn as shown dotted in Fig. 11j. The slope of this line may be measured to give:

$$\underline{\underline{L_n/V_n = 0·51}}$$

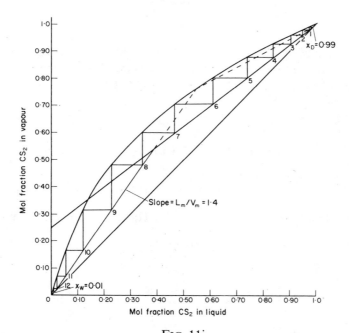

FIG. 11j

Problem 11.13

A fractionating column is required to distil a liquid containing 25% benzene and 75% toluene by weight so as to give a product of 90% benzene. A reflux ratio of 3·5 is to be used, and the feed will enter at its boiling-point.

If the plates used are 100% efficient, calculate by the Lewis–Sorel method the composition of liquid on the third plate, and by McCabe and Thiele's method estimate the number of plates required.

Solution

The equilibrium data for this problem are plotted as Fig. 11k. Converting wt% to mol fraction gives $x_f = 0·282$ and $x_d = 0·913$. There are no data given on the bottom product so that the normal mass balances cannot be applied. The equation of the top operating line is:

$$y_n = \frac{L_n}{V_n} x_{n+1} + \frac{D}{V_n} x_d \qquad \text{(equation 11.23)}$$

Now L_n/V_n is the slope of the top operating line which passes through the points (x_d, x_d) and $0, x_d/(R+1)$. This line is drawn in Fig. 11k and its slope measured or calculated as 0·78. The reflux ratio which is equal to L_n/D is given as 3·5, so that D/V_n may then be found since,

$$\frac{D}{V_n} = \frac{D}{L_n} = \frac{1}{3·5} \times 0·78 = 0·22$$

$$\therefore \qquad y_n = 0·78_{n+1} + 0·22x_d$$

$$= 0·78_{n+1} + 0·20$$

The composition of the vapour y_t leaving the top plate must be the same as the top product x_d since all the vapour is condensed. The composition of the liquid on the top plate x_t is found from the equilibrium curve since it is in equilibrium with vapour of composition $y_t = x_d = 0·913$.

$$\therefore \qquad x_t = 0·805$$

The composition of the vapour rising to the top plate y_{t-1} is found from the equation of the operating line:

$$y_{t-1} = 0·78 \times 0·805 + 0·20 = 0·828$$

x_{t-1} is in equilibrium with y_{t-1} and is found to equal 0·66 from the equilibrium curve.

Then $y_{t-2} = 0·78 \times 0·66 + 0·20 = 0·715$

Similarly $x_{t-2} = 0·50$

and $y_{t-3} = 0·78 \times 0·50 + 0·20 = 0·557$

when $\underline{\underline{x_{t-3} = 0·335}}$

The McCabe–Thiele construction in Fig. 11k shows that 5 theoretical plates are required in the rectifying section.

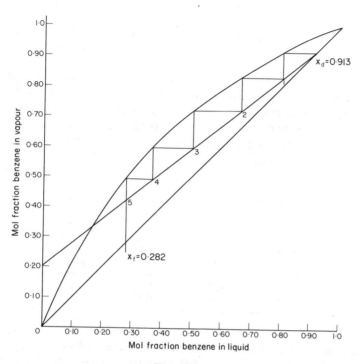

FIG. 11k

Problem 11.14

A 50 mol % mixture of benzene and toluene is fractionated in a batch still which has the separating power of 8 theoretical plates. It is proposed to obtain a constant quality product with a mol % benzene of 95, and to continue the distillation until the still has a content of 10 mol % of benzene.

What will be the range of reflux ratios used in the process?

Show graphically the relation between the required reflux ratio and the amount of distillate removed.

Solution

If a constant product is to be obtained from a batch still, the reflux ratio must be constantly increased. Initially S_1 kmol of liquor are in the still with a composition x_{S_1} of the MVC and a reflux ratio of R_1 is required to give the desired product composition x_d. Later, when S_2 kmol remain in the still of composition x_{S_2}, the reflux ratio has increased to R_2 when the amount of product is D kmol.

From an overall mass balance:

$$S_1 - S_2 = D$$

For the MVC:

$$S_1 x_{s_1} - S_2 x_{s_2} = D x_d$$

from which

$$D = S_1 \frac{(x_{s_1} - x_{s_2})}{(x_d - x_{s_2})} \qquad \text{(equation 11.75)}$$

In this problem $x_{s_1} = 0.5$ and $x_d = 0.95$ and these are 8 theoretical plates in the column. It remains, by using the equilibrium data, to determine values of x_{s_2} for selected reflux ratios. This is done graphically by choosing an intercept on the y-axis, calculating R, drawing the resulting operating line, and stepping off in the normal way 8 theoretical plates and finding the still composition x_{s_2} and hence D. The results of this process are shown below for $S_1 = 100\,\text{kmol}$.

$\phi = x_d/(R+1)$	R	x_{s_2}	D
0·4	1·375	0·48	4·2
0·35	1·71	0·405	17·3
0·30	2·17	0·335	26·8
0·25	2·80	0·265	34·3
0·20	3·75	0·195	40·3
0·15	5·33	0·130	45·1
0·10	8·50	0·090	47·7

The initial and final values of R are most easily determined by plotting R against x_{s_2} as shown in Fig. 111. The initial value of R corresponds to the initial still composition of 0·50 and is seen to be 1·3 and at the end of the process when $x_{s_2} = 0.1$, $R = 7.0$.

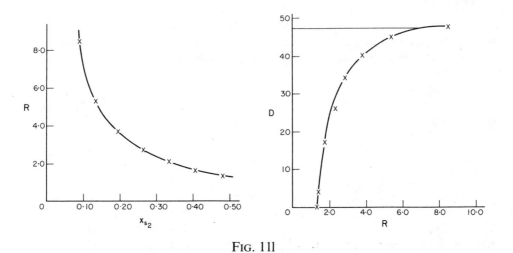

Fig. 111

Figure 111 includes a plot of reflux ratio against the quantity of distillate. When $R = 7.0$, $D = 47.1\,\text{kmol}/100\,\text{kmol}$ charged initially.

Problem 11.15

The vapour composition on a plate of a distillation column is:

$$
\begin{array}{lll}
C_1 & 0{\cdot}025 \text{ mol fraction} & 36{\cdot}5 \text{ rel. volatility} \\
C_2 & 0{\cdot}205 & 7{\cdot}4 \\
i-C_3 & 0{\cdot}210 & 3{\cdot}0 \\
n-C_3 & 0{\cdot}465 & 2{\cdot}7 \\
i-C_4 & 0{\cdot}045 & 1{\cdot}3 \\
n-C_4 & 0{\cdot}050 & 1{\cdot}0
\end{array}
$$

What will be the composition of the liquid on the plate if it is in equilibrium with the vapour?

Solution

In a mixture of A, B, C, D, etc., let the mol fraction in the liquid be x_A, x_B, x_C, x_D, etc., and in the vapour y_A, y_B, y_C, and y_D.

Now

$$x_A + x_B + x_C + \cdots = 1$$

$\therefore$

$$\frac{x_A}{x_B} + \frac{x_B}{x_B} + \frac{x_C}{x_B} + \cdots = \frac{1}{x_B}$$

But

$$\frac{x_A}{x_B} = \frac{y_A}{y_B \, \alpha_{AB}}$$

$\therefore$

$$\frac{y_A}{y_B \, \alpha_{AB}} + \frac{y_B}{y_B \, \alpha_{BB}} + \frac{y_C}{y_B \, \alpha_{CB}} + \cdots = \frac{1}{x_B}$$

or

$$\sum \left(\frac{y_A}{\alpha_{AB}} \right) = \frac{y_B}{x_B}$$

But

$$y_B = \frac{y_A \, x_B}{\alpha_{AB} \, x_A}$$

and substituting:

$$\sum \left(\frac{y_A}{\alpha_{AB}} \right) = \frac{y_A \, x_B}{\alpha_{AB} \, x_A \, x_B}$$

$\therefore$

$$x_A = \frac{(y_A/\alpha_{AB})}{\sum (y_A/\alpha_{AB})}$$

Similarly,

$$x_B = \frac{(y_B/\alpha_{BB})}{\sum (y_A/\alpha_{AB})} \quad \text{and} \quad x_C = \frac{(y_C/\alpha_{BC})}{\sum (y_A/\alpha_{AB})}$$

These relationships may be used to solve this problem and the calculation is best carried out in tabular form.

Component	y_i	α	y_i/α	$x_i = (y_i/\alpha)/\Sigma(y_i/\alpha)$
C_1	0·025	36·5	0·00068	0·002
C_2	0·205	7·4	0·0277	0·078
$i-C_3$	0·210	3·0	0·070	0·197
$n-C_3$	0·465	2·7	0·1722	0·485
$i-C_4$	0·045	1·3	0·0346	0·097
$n-C_4$	0·050	1·0	0·050	0·141
			$\Sigma(y_i/\alpha) = 0·355$	1·000

Problem 11.16

A liquor of 0·30 mol fraction of benzene and the rest toluene is fed to a continuous still to give a top product of 0·90 mol fraction benzene and a bottom product of 0·95 mol fraction toluene.

If the reflux ratio is 5·0, how many plates are required:

(a) If the feed is saturated vapour?
(b) If the feed is liquid at 283 K?

Solution

This problem is solved by the use of the McCabe–Thiele method which has been used in earlier problems. Here the q-line has two widely differing slopes and the effect of the feed condition is to alter the number of theoretical stages (Fig. 11m).

$$q = \frac{\lambda + H_{fs} - H_f}{\lambda}$$

where λ is the molar latent heat of vaporisation, H_{fs} is the molar enthalpy of the feed at its boiling-point, and H_f is the molar enthalpy of the feed.

For benzene and toluene: $\lambda = 30 \, \text{MJ/kmol}$

specific heat $= 1·84 \, \text{kJ/kg K}$

Boiling-points of benzene and toluene $= 353·3$ and $383·8$ K respectively.

(a) If the feed is a saturated vapour, $q = 0$.
(b) If the feed is a cold liquid at 283 K, the mean molecular weight is:

$$(0·3 \times 78) + (0·7 \times 92) = 87·8 \, \text{kg/kmol}$$

and the mean boiling-point $= (0·3 \times 353·3) \times (0·7 \times 383·8) = 374·7$ K.

Using a datum of 273 K:

$$H_{fs} = 1·84 \times 87·8(374·7 - 273) = 16{,}425 \, \text{kJ/kmol}$$
$$= 16·43 \, \text{MJ/kmol}$$

$$H_f = 1·84 \times 87·8(283 - 273) = 1615 \, \text{kJ/kmol}$$
$$= 1·615 \, \text{MJ/kmol}$$

$$\therefore \quad q = (30 + 16.43 - 1.615)/30 = 1.49$$

From equation 11.30, the slope of the q-line is $q/(q-1)$.
Hence slope $= 1.49/0.49 = 3.05$.

Thus for (a) and (b) the slope of the q-line is zero and 3.05 respectively, and in Fig. 11m these lines are drawn through the point (x_f, x_f).

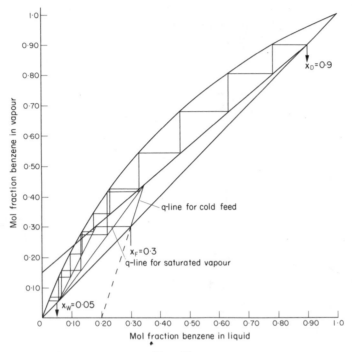

FIG. 11m

By stepping off the ideal stages, the following results are obtained:

Feed	Theoretical plates		
	Stripping section	Rectifying section	Total
Saturated vapour	4	5	9
Cold liquid	4	3	7

Thus a cold feed requires fewer plates than a vapour feed but the capital cost saving is offset by the increased heat load on the reboiler.

Problem 11.17

A mixture of alcohol and water with 0·45 mol fraction of alcohol is to be continuously distilled in a column so as to give a top product of 0·825 mol fraction alcohol and a liquor at the bottom with 0·05 mol fraction alcohol.

How many theoretical plates are required if the reflux ratio used is 3? Indicate on a diagram what is meant by the Murphree plate efficiency.

Solution

This example is solved by a simple application of the McCabe–Thiele method and is illustrated in Fig. 11n, where it is seen that 10 theoretical plates are required. Murphree plate efficiency is discussed in the solution to Problem 11.9.

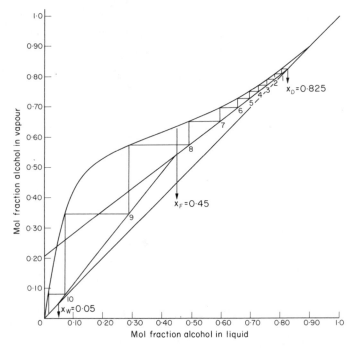

$x_D = 0.825$

$x_F = 0.45$

$x_W = 0.05$

Mol fraction alcohol in vapour

Mol fraction alcohol in liquid

FIG. 11n

Problem 11.18

It is required to separate 1 kg/s of a solution of ammonia in water, containing 30% by weight of ammonia, to give a top product of 99·5% purity and a weak solution containing 10% by weight of ammonia.

Calculate the heat required in the boiler and the heat to be rejected in the condenser, assuming a reflux 8% in excess of the minimum and a column pressure of 1013 kN/m². The plates may be assumed to have an ideal efficiency of 60%.

Solution

Taking a material balance for the whole throughput and for the ammonia gives:

$$D' + W' = 1.0$$

$$0.995D' + 0.1W' = 1.0 \times 0.3$$

Thus $$D' = 0{\cdot}22 \text{ kg/s}$$

and $$W' = 0{\cdot}78 \text{ kg/s}$$

The enthalpy composition chart for this system is shown in Fig. 11o. It is assumed that the feed F and the bottom product W are liquids at their boiling-points.

Location of the poles N and M

N_m for minimum reflux will be found by drawing a tie line through F, representing the feed, to cut the line $x = 0{\cdot}995$ at N_m.

The minimum reflux ratio

$$R_m = \frac{\text{length } N_m A}{\text{length } AL} = \frac{1952 - 1547}{1547 - 295} = 0{\cdot}323$$

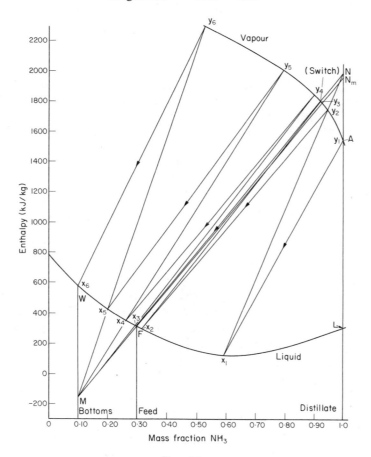

FIG. 11o

Since the actual reflux is 8% above the minimum,

$$NA = 1{\cdot}08 N_m A$$

$$= 1{\cdot}08 \times 405 = 437$$

Point N therefore has an ordinate $437 + 1547 = 1984$ and abscissa 0·995.

Point M is found by drawing NF to cut the line $x = 0·10$, through W, at M.

Then the number of theoretical plates is found, as on the diagram, to be $5+$.

The number of plates to be provided $= 5/0·6 = 8·33$, say 9.

The feed is introduced just below the third ideal plate from the top, or just below the fifth actual plate.

Heat input at the boiler per unit mass of bottom product

$$= Q_B/W = 582 - (-209) = 791 \text{ kJ/kg} \quad \text{(from equation 11.62).}$$

Heat input to boiler $= 791 \times 0·78 = \underline{617 \text{ kW.}}$

Condenser duty $=$ length $NL \times D$
$$= (1984 - 296) \times 0·22 = \underline{372 \text{ kW.}}$$

Problem 11.19

A mixture of 60 mol % benzene, 30% of toluene, and 10% xylene is run into a batch still. If the top product is to be 99% benzene, determine:

(a) the liquid composition on each plate at total reflux;
(b) the composition on the second and fourth plates for $R = 1·5$;
(c) as for (b) but $R = 3$;
(d) as for (c) but $R = 5$;
(e) as for (d) but $R = 8$ and for the conditions when the mol % benzene in the still is 10;
(f) as for (e) but with $R = 5$.

Take the relative volatility of benzene to toluene as 2·4 and xylene to toluene as 0·43.

Solution

Although this problem is one of multicomponent batch distillation, the product remains of constant composition so that normal methods can be used for plate-to-plate calculations at a point value of the varying reflux ratio.

(a) At total reflux, for each component the operating line is

$$y_n = x_{n+1}$$

Also
$$y = \alpha x / \Sigma(\alpha x)$$

Setting the solution in tabular form:

	α	x_s	αx_s	$y_s = x_1$	αx_1	$y_1 = x_2$	αx_2	$y_2 = x_3$	Similarly x_4	x_5
B	2·4	0·60	1·44	0·808	1·164	0·867	2·081	0·942	0·975	0·989
T	1·0	0·30	0·30	0·168	0·168	0·125	0·125	0·057	0·025	0·011
X	0·43	0·10	0·043	0·024	0·010	0·008	0·003	0·001	—	—
			1·783	1·000	1·342	1·000	2·209	1·000	1·000	1·000

(b) The operating line for the rectifying section is:

$$y_n = \frac{L_n}{V_n}x_{n+1} + \frac{D}{V_n}x_d$$

Now

$$R = L_n/D \quad \text{and} \quad V = L_n + D$$

$$\therefore \qquad y_n = \frac{R}{R+1}x_{n+1} + \frac{x_d}{R+1}$$

If $R = 1\cdot5$:

for benzene: $y_{nb} = 0\cdot6x_{n+1} + 0\cdot396$

toluene: $y_{nt} = 0\cdot6x_{n+1} + 0\cdot004$

xylene: $y_{nx} = 0\cdot6x_{n+1}$

The liquid composition on each plate is then found from these operating lines.

	y_s	x_1	αx_1	y_1	x_2	αx_2	y_2	x_3	αx_3	y_3	x_4
B	0·808	0·687	1·649	0·850	0·757	1·817	0·848	0·754	1·810	0·899	0·838
T	0·168	0·273	0·273	0·141	0·228	0·228	0·107	0·171	0·171	0·085	0·135
X	0·024	0·040	0·017	0·009	0·015	0·098	0·045	0·075	0·032	0·016	0·027
	1·000	1·000	1·939	1·000	1·000	2·143	1·000	1·000	2·013	1·000	1·000

(c) If $R = 5$, the operating line equations become:

$$y_{nb} = 0\cdot833x_{n+1} + 0\cdot165$$

$$y_{nt} = 0\cdot833x_{n+1} + 0\cdot0017$$

$$y_{nx} = 0\cdot833x_{n+1}$$

	y_s	x_1	αx_1	y_1	x_2	αx_2	y_2	x_3	αx_3	y_3	x_4
B	0·808	0·772	1·853	0·897	0·879	2·110	0·947	0·939	2·254	0·974	0·971
T	0·168	0·200	0·200	0·097	0·114	0·114	0·051	0·059	0·059	0·025	0·028
X	0·024	0·028	0·012	0·006	0·007	0·003	0·002	0·002	0·001	0·001	0·001
	1·000	1·000	2·065	1·000	1·000	2·227	1·000	1·000	2·314	1·000	1·000

(d) When the benzene content is 10% in the still, a mass balance gives the mols of distillate removed as (assume 100 kmol initially):

$$D = 100(0\cdot6 - 0\cdot1)/(0\cdot99 - 0\cdot1)$$

$$= 56\cdot2 \text{ kmol removed}$$

$\therefore$ 43·8 kmol remain of which 4·38 are benzene, $x_b = 0\cdot10$

29·42 are toluene, $x_t = 0\cdot67$

10·00 are xylene, $x_x = 0\cdot23$

If $R = 8$, the operating lines become:

$$y_{nb} = 0{\cdot}889x_{n+1} + 0{\cdot}11$$

$$y_{nt} = 0{\cdot}889x_{n+1} + 0{\cdot}001$$

$$y_{nx} = 0{\cdot}889x_{n+1}$$

	x_s	αx_s	y_s	x_1	αx_1	y_1	x_2	αx_2	y_2	x_3	αx_3	y_3	x_4
B	0·10	0·24	0·24	0·146	0·350	0·307	0·222	0·533	0·415	0·343	0·823	0·560	0·506
T	0·67	0·67	0·66	0·741	0·741	0·650	0·730	0·730	0·569	0·639	0·639	0·435	0·488
X	0·23	0·10	0·10	0·113	0·049	0·043	0·048	0·021	0·016	0·018	0·008	0·005	0·006
		1·01	1·00	1·000	1·140	1·000	1·000	1·284	1·000	1·000	1·470	1·000	1·000

(f) Exactly the same procedure is repeated for this part of the question, when the operating lines become:

$$y_{nb} = 0{\cdot}833x_{n+1} + 0{\cdot}165$$

$$y_{nt} = 0{\cdot}833x_{n+1} + 0{\cdot}0017$$

$$y_{nx} = 0{\cdot}833x_{n+1}$$

Problem 11.20

A continuous still is fed with a mixture of mol fraction 0·5 of the more volatile component, and gives a top product of 0·9 mol fraction more volatile component and a bottom product of 0·10 mol fraction.

If the still operates with an L_n/D ratio of 3·5:1, calculate by Sorel's method the composition of the liquid on the third theoretical plate from the top:

(a) for benzene–toluene; and
(b) for n-heptane–toluene.

Solution

A series of mass balances as described in previous problems enables the flows within the column to be calculated as:

For a basis of 100 kmol of feed and a reflux ratio of 3·5:

$$D = 50, \quad L_n = 175, \quad V_n = 225 \text{ kmol}$$

The top operating line equation is then:

$$y_n = 0{\cdot}778x_{n+1} + 0{\cdot}20$$

(a) Use will be made of the equilibrium data from earlier examples involving benzene and toluene.

The vapour leaving the top plate has the composition as the top product, i.e. $y_t = 0{\cdot}90$. From the equilibrium data, $x_t = 0{\cdot}78$.

Then $\quad\quad\quad\quad y_{t-1} = 0{\cdot}778 \times 0{\cdot}78 + 0{\cdot}20 = 0{\cdot}807$

and $\quad\quad\quad\quad x_{t-1}$ from equilibrium data $= 0{\cdot}640$

Similarly, $\quad\quad\quad y_{t-2} = 0{\cdot}778 \times 0{\cdot}640 + 0{\cdot}20 = 0{\cdot}698$

$\quad\quad\quad\quad\quad\quad x_{t-2} = 0{\cdot}49$

$\quad\quad\quad\quad\quad\quad y_{t-3} = 0{\cdot}778 \times 0{\cdot}490 + 0{\cdot}20 = 0{\cdot}581$

and $\quad\quad\quad\quad \underline{\underline{x_{t-3} = 0{\cdot}36}}$

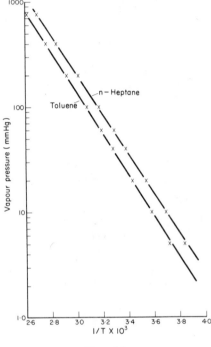

Fig. 11p

(b) Vapour pressure data from Perry[1] for n-heptane and toluene are plotted in Fig. 11p. These data may be used to calculate an average value of the relative volatility α from:

$$\alpha = P_H^0/P_T^0$$

T	$10^3 T$	P_H^0	P_T^0	α	
370	2·7	730	510	1·43	
333	3·0	200	135	1·48	Average $\alpha = 1{\cdot}52$
303	3·3	55	35	1·57	
278	3·6	15	9·4	1·60	

[1] PERRY, J. H.: *Chemical Engineers' Handbook*, 5th edn. (McGraw-Hill, 1973). Revised by R. PERRY and C. CHILTON.

Instead of drawing the equilibrium curve for the system, point values may be calculated from:

$$x = \frac{y}{\alpha - (\alpha - 1)\, y} \qquad \text{(equation 11.9)}$$

Then $y_t = 0.9, \quad x_t = 0.9/(1.52 - 0.52 \times 0.9) = 0.856$

From the same operating line equation:

$$y_{t-1} = 0.778 \times 0.856 + 0.20 = 0.865$$

Similarly $x_{t-1} = 0.808$

$$y_{t-2} = 0.829$$

$$x_{t-2} = 0.761$$

$$y_{t-3} = 0.792$$

and $\underline{\underline{x_{t-3} = 0.715}}$

Problem 11.21

A mixture of 40 mol % benzene with toluene is distilled in a column to give a product of 95 mol % benzene and a waste of 5 mol % benzene, using a reflux ratio of 4.

(a) Calculate by Sorel's method the composition on the second plate from the top.
(b) Using McCabe and Thiele's method, determine the number of plates required and the position of the feed if supplied to the column as liquid at the boiling-point.
(c) Find the minimum reflux ratio possible.
(d) Find the minimum number of plates.
(e) If the feed is passed in at 288 K find the number of plates required using the same reflux ratio.

Solution

The equilibrium data for benzene and toluene are plotted in Fig. 11q.

Now $x_f = 0.40, \quad x_d = 0.95, \quad x_w = 0.05, \quad \text{and} \quad R = 4.0$

so that mass balances may be performed to calculate the operating line equations. Taking a basis of 100 kmol:

$$100 = D + W$$

$$100 \times 0.4 = 0.95D + 0.05W$$

from which $D = 38.9 \quad \text{and} \quad W = 61.1 \, \text{kmol}$

Now $L_n/D = 4 \quad \text{so that} \quad L_n = 155.6 \, \text{kmol}$

$$V_n = L_n + D = 194.5 \, \text{kmol}$$

(a) The top operating line equation is:

$$y_n = \frac{L_n}{V_n} x_{n+1} + \frac{D}{V_n} x_d \qquad \text{(equation 11.19)}$$

i.e.
$$y_n = (155{\cdot}6/194{\cdot}5)\, x_{n+1} + (38{\cdot}9/194{\cdot}5) \times 0{\cdot}95$$

$$y_n = 0{\cdot}8 x_{n+1} + 0{\cdot}19$$

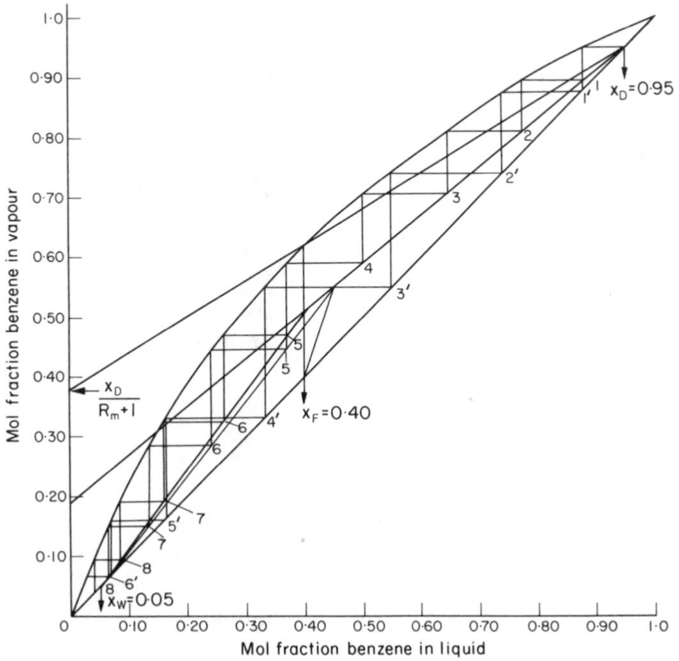

FIG. 11q

Vapour y_t leaving the top tray has a composition $= x_d = 0{\cdot}95$ so that x_t, the liquid composition on the top tray, is found from the equilibrium curve to be equal to $0{\cdot}88$. y_{t-1} is found from the operating line equation as:

$$y_{t-1} = 0{\cdot}8 \times 0{\cdot}88 + 0{\cdot}19 = 0{\cdot}894$$

$$x_{t-1} = \text{from the equilibrium curve} = 0{\cdot}775$$

$$\therefore \qquad y_{t-2} = 0{\cdot}8 \times 0{\cdot}775 + 0{\cdot}19 = 0{\cdot}810$$

$$x_{t-2} = \text{from the equilibrium curve} = \underline{0{\cdot}645}$$

(b) The steps in the McCabe–Thiele determination are shown in Fig. 11q where 8 theoretical plates are required with a boiling liquid feed.

(c) The minimum reflux ratio corresponds to an infinite number of plates. This condition occurs when the top operating line passes through the intersection of the *q*-line and the equilibrium curve. This line is seen to give an intersection on the *y*-axis equal to 0·375.

$$\therefore \qquad 0·375 = x_d/(R_M + 1)$$

and

$$R_M = 1·53$$

(d) The minimum number of plates occurs at total reflux and may be determined by stepping between the equilibrium curve and the diagonal $y = x$ to give 6 theoretical plates as shown.

Alternatively, Fenske's equation may be used.

$$n + 1 = \log \frac{[(x_A/x_B)_d(x_B/x_A)_s]}{\log \alpha_{AB}} \qquad \text{(equation 11.42)}$$

$$= \frac{\log (0·95/0·05)(0·95/0·05)}{\log 2·4}$$

and

$$n = 5·7 \quad \text{or} \quad \underline{\underline{6 \text{ plates}}}$$

(e) If a cold feed is introduced, the *q*-line is no longer vertical. Its slope may be calculated as shown in Problem 11.16. In this problem, *q* is found to be 1·45 and the *q*-line has a slope of 3·22. This line is shown in Fig. 11q and the number of theoretical plates is found to be unchanged at $\underline{\underline{8}}$.

Problem 11.22

7·5 kg/s of n-propanol is obtained off the top of a distillation column working at 205 kN/m². The propanol is totally condensed and part taken for product and part returned for reflux. Under these conditions the boiling-point is 390 K and the liquid density 775 kg/m³.

Determine the diameter and plate spacing based on entrainment and estimate the height of liquor in the downcomer. The liquor flow down may be taken as 5·0 kg/s.

What size do you suggest for the downcomer, the weir, the liquid seal, and the area of the risers to the bubble caps? The F_w factor for the height over the weir may be taken as 1.

Solution

Vapour flow rate = 7·5 kg/s.
Liquid flow rate = 5·0 kg/s.
Molecular weight of propanol = 60 kg/kmol.
At 205 kN/m² and 390 K, the density of propanol vapour is given by:

$$\rho_v = (60/22·4)(273/290)(205/101.3) = 3·79 \text{ kg/m}^3$$

$$\frac{L}{G}\sqrt{\left(\frac{\rho_v}{\rho_L}\right)} = \frac{5·0}{7·5}\sqrt{\left(\frac{3·79}{775}\right)} = 0·047$$

If a plate spacing of 0·46 m (18 in) is assumed, Fig. 11.57 may be used to find the parameter C_{sb} as 0·10 m/s.

$$0·10 = u_n \sqrt{(\rho_v/\rho_L - \rho_v)}$$

and $u_n = 1·42$ m/s

This is the velocity at which flooding will occur and the normal maximum operating velocity of 80% of this value will be taken, i.e. 1·14 m/s. If a single-pass crossflow tray is used, the net area of the tray will be 88% of the tower area, A_t.

Volumetric vapour flow = 7·5/3·79 = 1·98 m³/s.

$$A_t = 1·98/0·88 \times 1·14 = 1·97 \text{m}^2$$

and $\underline{D = 1·58 \text{ m}}$

This is the first estimate of diameter which must now be checked with the hydraulic parameters.

From reference 46, Chap. 11, Vol. 2 for a tray of this diameter, 76 mm (3 in) caps would be used.

Slot area = 3065 mm².   Plate area/cap = 10,130 mm².
Riser area = 2000 mm².   Hole pitch/diameter = 2·13.
Pitch = 108 mm.   Free area of risers = 20%.

The maximum liquid flow normally allowed = 0·025 m³/s m (10 US gpm/in)

$$= 25 \text{ kg/s m}$$

Actual liquid flow rate = 5·0 kg/s.
Weir length for single-pass crossflow tray = 0·77D_t = 1·22 m (48 in).

$$\text{liquid flow} = 5·0/1·22 = 4·1 \text{ kg/s m}$$

which is acceptable.

The depth of liquid in the downcomer Z'_p is given by:

$$Z'_p = 2(h_w + h_{ow}) + h_d + h_{rc} \qquad \text{(equation 11.109)}$$

where h_w is the weir height, h_{ow} is the height of crest over the weir, h_d is the head loss at bottom of downcomer, and h_{rc} is the head loss due to riser and reversal inside cap.

The height of the weir h_w may be taken as 64 mm (Table 11.4, Vol. 2).

$$h_{ow} = 68,175 F_w (Q'w)^{2/3} \qquad \text{(equation 11.106)}$$

where $F_w = 1$ (given), Q' is the liquid flow rate = 5·0/775 = 0·0065 m³/s, and w is the weir length = 1·22 m ≡ 1220 mm.

$$h_{ow} = 68,175 \times 1(0·0065/1220)^{2/3} = 20·7 \text{ mm}$$

Now $h_d = 13(Q'/A_{d_a})^2$

where $Q' = 0·0065$ m³/s and A_{d_a} is the area under downcomer apron which for a 38 mm apron clearance = 0·038 × 1·22 = 0·046 m².

$$h_d = 131(0·0065/0·046)^2 = 2·62 \text{ mm}$$

h_{rc} is given by: $h_{rc} = 301u_r^2 \rho_v/\rho_L$

$$A_r = 0.2 \times 1.97 = 0.39 \text{ m}^2$$

$$u_r = 1.98/0.39 = 5.08 \text{ m/s}$$

$$\therefore \quad h_{rc} = 301(5.08)^2 \times 3.79/775$$

$$= 38.0 \text{ mm}$$

$$\therefore \quad Z'_p = 2(64 + 20.7) + 2.62 + 38.0$$

$$= 210 \text{ mm} = 0.21 \text{ m}$$

The actual tray spacing should be at least twice this value (say 0·4 m). As the diameter was calculated on an assumed spacing of 0·46 m, this may be taken as satisfactory and hence:

$$D_t = 1.58 \text{ m} \quad \text{Spacing} = 0.46 \text{ m}$$

Problem 11.23

Determine the minimum reflux ratio by the following two methods for the three systems given below:

(a) using Underwood's equation; and
(b) using Colburn's rigorous method.

1. C_6 (0·60 mol fraction) C_7 (0·30) C_8 (0·10) to give a product of 0·99 mol fraction C_6.

	Mol frac.	Rel. vol.	x_d
2. Components A	0·3	2	1·0
B	0·3	1	—
C	0·4	$\frac{1}{2}$	—
3. Components A	0·25	2	1·0
B	0·25	1	—
C	0·25	$\frac{1}{2}$	—
D	0·25	$\frac{1}{4}$	—

Solution

(a) Under conditions where the relative volatility remains constant, Underwood developed the following equations from which R_m may be calculated:

$$\frac{\alpha_A x_{fA}}{\alpha_A - \theta} + \frac{\alpha_B x_{fB}}{\alpha_B - \theta} + \frac{\alpha_C x_{fC}}{\alpha_C - \theta} + \cdots = 1 - q \qquad \text{(equation 11.91)}$$

and

$$\frac{\alpha_A x_{dA}}{\alpha_A - \theta} + \frac{\alpha_B x_{dB}}{\alpha_B - \theta} + \frac{\alpha_C x_{dC}}{\alpha_C - \theta} + \cdots = R_m + 1 \qquad \text{(equation 11.92)}$$

Where x_{fA}, x_{fB}, x_{dA}, x_{dB}, etc., are the mol fractions of components A and B, etc., in the feed and distillate with A the light key and B the heavy key.

α_A, α_B, α_C, etc., are the volatilities with respect to the least volatile component.

θ is the root of equation 11.91 and lies between the values of α_A and α_B. Thus θ may be calculated from equation 11.91 and substituted into 11.92 to give R_m.

(b) Colburn's method allows the value of R_m to be calculated from approximate values of the pinch compositions of the key components. This value may then be checked against empirical relationships as shown in a worked example in Vol. 2.

The method is long and tedious and only the first approximation will be worked here.

$$R_m = \frac{1}{\alpha_{AB} - 1}\left[\left(\frac{x_{dA}}{x_{nA}}\right) - \alpha_{AB}\left(\frac{x_{dB}}{x_{nB}}\right)\right] \qquad \text{(equation 11.85)}$$

Where x_{dA} and x_{nA} are the top and pinch compositions of the light key component, x_{dB} and x_{nB} are the top and pinch compositions of the heavy key component, and α_{AB} is the volatility of the light key relative to the heavy key component.

The difficulty in using this equation is that the values of x_{nA} and x_{nB} are only known in special cases where the pinch coincides with the feed composition. Colburn has suggested that an approximate value for x_{nA} is given by:

$$x_{nA}\,(\text{approx.}) = \frac{r_f}{(1 + r_f)(1 + \Sigma\alpha x_{fh})} \qquad \text{(equation 11.86)}$$

and

$$x_{nB}\,(\text{approx.}) = \frac{x_{nA}}{r_f} \qquad \text{(equation 11.87)}$$

where r_f is the estimated ratio of the key components on the feed plate.

For an all liquid feed at its boiling-point, r_f equals the ratio of the key components in the feed. Otherwise r_f is the ratio of the key components in the liquid part of the feed, x_{fh} is the mol fraction of each component in the liquid portion of feed heavier than the heavy key, and α is the volatility of the component relative to the heavy key.

(1) Relative volatility data are required and it will be assumed that $\alpha_{68} = 5$, $\alpha_{67} = 2\cdot5$, and also that $x_{d7} = 0\cdot01$ and that $q = 1$.

Then substituting in Underwood's equation gives:

$$\frac{5 \times 0\cdot60}{5 - \theta} + \frac{2\cdot5 \times 0\cdot30}{2\cdot5 - \theta} + \frac{0\cdot1 \times 1}{1 - \theta} = 1 - q = 0$$

From which by trial and error, $\theta = 3\cdot1$.

Then:
$$\frac{5 \times 0\cdot99}{5 - 3\cdot1} + \frac{2\cdot5 \times 0\cdot01}{2\cdot5 - 3\cdot1} = R_m + 1$$

and
$$R_m = 1\cdot57$$

In Colburn's equation using C_6 and C_7 as the light and heavy keys respectively:

$$r_f = 0\cdot6/0\cdot3 = 2\cdot0$$

$$\Sigma\alpha x_{fh} = (1/2\cdot5) \times 0\cdot1 = 0\cdot04$$

$$x_{nA} = 2/(3 \times 1\cdot04) = 0\cdot641$$

$$\alpha_{AB} = 5\cdot0/2\cdot5 = 2\cdot0$$

$$x_{nB} = 0\cdot641/2 = 0\cdot32$$

$$Rm = \frac{1}{1}\left(\frac{0.99}{0.641} - \frac{2 \times 0.01}{0.32}\right)$$

and
$$R_m = 1.48$$

(2) The light key = A, heavy key = B, and $\alpha_A = 4$, $\alpha_B = 2$, $\alpha_C = 1$, and if q is assumed to equal 1, then substitution into Underwood's equations gives $\theta = 2.8$ and $R_m = 2.33$. In Colburn's method, $x_{dB} = 0$ and $r_f = 0.3/0.3 = 1.0$.

$$\Sigma \alpha n_{fh} = (1/2) \times 0.4 = 0.2$$

$$x_{nA} = 1/(2 \times 1.2) = 0.417$$

$$\alpha_{AB} = 4/2 = 2.0$$

and
$$R_m = 2.40$$

(3) In this case $\alpha_A = 8$, $\alpha_B = 4$, $\alpha_C = 2$, $\alpha_D = 1$. If $q = 1$, then θ is found from Underwood's equation to equal 5.6 and $R_m = 2.33$.

In Colburn's method, $x_{dB} = 0$ and $r_f = 1.0$, $LK = A$, $HK = B$.

$$\Sigma \alpha x_{fh} = (0.5 \times 0.25) + (0.25 \times 0.25) = 0.188$$

$$x_{nA} = 1/(2 \times 1.188) = 0.421$$

$$\alpha_{AB} = 2$$

and
$$R_m = 2.38$$

In all cases, good agreement is shown between the two methods.

Problem 11.24

A liquor consisting of phenol and cresols with some xylenols is fractionated to give a top product of 95.3 mol % phenol. The compositions of the top product and of the phenol in the bottoms are given. A reflux ratio of 10 will be used.

(a) Complete the material balance over the still for a feed of 100 kmol.
(b) Calculate the composition on the second plate from the top.
(c) Calculate the composition on the second plate from the bottom.
(d) Calculate the minimum reflux ratio by Underwood's equation and by Colburn's approximation.

	Compositions in mol %		
	Feed	Top	Bottom
Phenol	35	95.3	5.24
o-Cresol	15	4.55	
m-Cresol	30	0.15	
Xylenols	20		
	100	100	

Heavy key is m-cresol; light key is phenol.

Solution

(a) A mass balance overall and on phenol gives, with a basis of 100 kmol:

$$100 = D + W$$

$$100 \times 0.35 = 0.953D + 0.0524W$$

from which $\quad D = 33.0 \, \text{kmol} \quad$ and $\quad W = 67.0 \, \text{kmol}$

Balances on the remaining components give the required bottom products.

o-cresol: $\qquad 100 \times 0.15 = 0.0455 \times 33 + 67x_{wo}$

and $\qquad x_{wo} = \underline{0.2017}$

m-cresol: $\qquad 100 \times 0.30 = 0.0015 \times 33 + 67x_{wm}$

and $\qquad x_{wm} = \underline{0.4472}$

xylenols: $\qquad 100 \times 0.20 = 0 + 67x_{wx}$

and $\qquad x_{wx} = \underline{0.2987}$

(b)

$$L_n/D = 10 \quad \therefore L_n = 330 \, \text{kmol}$$

$$V_n = L_n + D \quad \text{and} \quad V_n = 363 \, \text{kmol}$$

The equation of the top operating line is:

$$y_n = \frac{L_n}{V_n} x_{n+1} + \frac{D}{V_n} x_d$$

$$= 0.91 x_{n+1} + 0.091 x_d$$

The operating lines for each component then become:

$$\text{P} \quad y_{np} = 0.91 x_{n+1} + 0.0867$$

$$\text{O} \quad y_{no} = 0.91 x_{n+1} + 0.0414$$

$$\text{M} \quad y_{nm} = 0.91 x_{n+1} + 0.000\,14$$

$$\text{X} \quad y_{nx} = 0.91 x_{n+1}$$

Average α-values will be taken from data given in Vol. 2 as:

$$\alpha_{PO} = 1.25, \quad \alpha_{OO} = 1.0, \quad \alpha_{MO} = 0.63, \quad \alpha_{XO} = 0.37$$

The solution is best set out as a table as follows, using the operating line equations and the equation

$$x = \frac{y/\alpha}{\Sigma(y/\alpha)}$$

	$y = x_d$	y_t/α	x_t	y_{t-1}	y_{t-1}/α	x_{t-1}
P	0·953	0·762	0·941	0·943	0·754	0·928
O	0·0455	0·0455	0·056	0·054	0·054	0·066
M	0·0015	0·0024	0·003	0·003	0·005	0·006
X	—	—	—	—	—	—
	$\Sigma(y_t/\alpha) = 0.8099$		1·000	1·000	0·813	1·000

(c) In the bottom of the column:

$$L_m = L_n + F = 430\,\text{kmol}$$
$$V_m = L_m - W = 363\,\text{kmol}$$

and
$$y_m = \frac{L_m}{V_m}x_{n+1} - \frac{W}{V_m}x_w$$
$$= 1.185x_{m+1} - 0.185x_w$$

Hence for each component:

$$P \quad y_{mp} = 1.185x_{m+1} - 0.0097$$
$$O \quad y_{mo} = 1.185x_{m+1} - 0.0373$$
$$M \quad y_{mm} = 1.185x_{m+1} - 0.0827$$
$$X \quad y_{mx} = 1.185x_{m+1} - 0.0553$$

Now using these operating lines to calculate x and also $y = \alpha x/\Sigma\alpha x$:

	x_s	αx_s	y_s	x_1	αx_1	y_1	x_2
P	0·0524	0·066	0·100	0·093	0·116	0·156	0·140
O	0·2017	0·202	0·305	0·289	0·289	0·387	0·358
M	0·4472	0·282	0·427	0·430	0·271	0·363	0·376
X	0·2987	0·111	0·168	0·188	0·070	0·094	0·126
	1·000 $\Sigma = 0.661$		1·000	1·000 $\Sigma = 0.746$		1·000	1·000

(d) Underwood's equations defined in the previous problem are used with $\alpha_p = 3.4$, $\alpha_o = 2.7$, $\alpha_m = 1.7$, $\alpha_x = 1.0$ to give:

$$\frac{3.4 \times 0.35}{3.4 - \theta} + \frac{2.7 \times 0.15}{2.7 - \theta} + \frac{1.7 \times 0.30}{1.7 - \theta} + \frac{1.0 \times 0.20}{1 - \theta} = 1 - q = 0$$

$3.4 > \theta > 1.7$ and θ is found by trial and error to equal 2·06.

Then:
$$\frac{3.4 \times 0.953}{3.4 - 2.06} + \frac{2.7 \times 0.0455}{2.7 - 2.06} + \frac{1.7 \times 0.0015}{1.7 - 2.06} = R_{m+1}$$

and
$$R_m = 1.60$$

Colburn's equation states:

$$R_m = \frac{1}{\alpha_{AB} - 1}\left[\left(\frac{x_{dA}}{x_{nA}}\right) - \alpha_{AB}\left(\frac{x_{dB}}{x_{nB}}\right)\right]$$

$$x_{nA} = \frac{r_f}{(1 + r_f)(1 + \Sigma x_{fh})}$$

$$x_{nB} = x_{nA}/r_f$$

A and B are the light and heavy keys, i.e. phenol and m-cresol.

$$r_f = 0.35/0.30 = 1.17$$

$$\Sigma \alpha x_{fh} = (0.37/0.63) \times 0.20 = 0.117$$

$$x_{nA} = 1.17/(2.17 \times 1.117) = 0.482$$

$$x_{nB} = 0.482/1.17 = 0.413$$

$$\alpha_{AB} = 1.25/0.63 = 1.98$$

$$\therefore \qquad R_m = \frac{1}{0.98}\left(\frac{0.953}{0.482} - \frac{1.98 \times 0.0015}{0.413}\right)$$

$$= \underline{\underline{1.95}}$$

This is the first approximation by Colburn's method and provides a good estimate of R_m.

Problem 11.25

A continuous fractionating column is to be designed to separate 2·5 kg/s of a mixture of 60% toluene and 40% benzene so as to give an overhead of 97% benzene and a waste of 98% toluene by weight.

A reflux ratio of 3·5 kmol of reflux per kmol of product is to be used and the molar latent heat of benzene and toluene may be taken as 30 MJ/kmol.

Calculate

(a) The weight of product and waste per unit time.
(b) The number of theoretical plates and position of feed if the feed is liquid at 295 K, of specific heat 1·84 kJ/kg K.
(c) How much steam at 240 kN/m² is required in the still.
(d) What will be the required diameter of the column if it operates at atmospheric pressure and a vapour velocity of 1·0 m/s.
(e) If the vapour velocity is to be 0·75 m/s, based on free area of column, the necessary diameter of the column.
(f) The minimum possible reflux ratio and the minimum number of plates for a feed entering at its boiling-point.

Solution

(a) An overall mass balance and a benzene balance allow the weight of product and waste to be calculated directly.

$$2.5 = D' + W'$$

$$2.5 \times 0.4 = 0.97D' + 0.02W'$$

from which $$W' = 1.5 \,\text{kg/s}$$

and $$D' = 1.0 \,\text{kg/s}$$

(b) This part of the problem will be solved by the McCabe–Thiele method. If the given compositions are converted to mol fractions, then:

$$x_f = 0.44, \quad x_w = 0.024, \quad x_d = 0.974$$

and a mass balance gives for a basis of 100 kmol of feed,

$$100 = D + W$$

$$100 \times 0.44 = 0.974S + 0.024D$$

from which $D = 43.8$ and $W = 56.2$ kmol per 100 kmol of feed

If $R = 3.5$, $L_n/D = 3.5$ and $L_n = 153.3$ kmol

$$V_n = L_n + D \quad \text{and} \quad V_n = 197.1 \,\text{kmol}$$

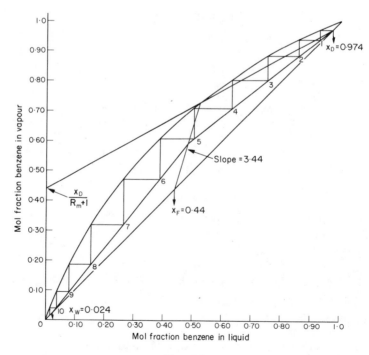

FIG. 11r

The intercept on the y-axis $= x_d/(R+1) = 0.216$.

As the feed is a cold liquid, the slope of the q-line must be found. Using the given data and employing the method used in earlier problems, q is found to be 1·41 and the slope $= q/(q+1) = 3.44$. This enables the diagram to be completed as shown in Fig. 11r from which it is seen that 10 theoretical plates are required with the feed tray as fifth from the top.

(c) The boil-up rate at the bottom of the column $= V_m$.

$$V_m = 238.1 \text{ kmol per } 100 \text{ kmol feed}$$

$$\text{feed rate} = 2.5/\text{mean mol wt.} = 2.5/86.4 = 0.0289 \text{ kmol/s}$$

$$\therefore \qquad \text{vapour rate} = (238.1/100) \times 0.0289$$

$$= 0.069 \text{ kmol/s}$$

$$\text{Heat load} = 0.069 \times 30 = 2.07 \text{ MW}$$

$$= 2070 \text{ kW}$$

Latent heat of steam at $240 \text{ kN/m}^2 = 2186 \text{ kJ/kg}$ (from Vol. 1)

$$\therefore \qquad \text{steam required} = 2070/2186 = 0.95 \text{ kg/s}$$

(d) At the top of the column the temperature is the boiling-point of essentially pure benzene, i.e. 353·3 K.

$$\therefore \qquad \rho = (1/22.4)(273/353.3) = 0.034 \text{ kmol/m}^3$$

$$V_n = 197.1 \text{ kmol per } 100 \text{ kmol of feed}$$

$$\text{Vapour flow} = (197.1/100) \times 0.0289 = 0.057 \text{ kmol/s}$$

$$\therefore \qquad \text{volumetric flow rate} = 0.057/0.034 = 1.68 \text{ m}^3/\text{s}$$

If vapour velocity $= 1.0 \text{ m/s}$,

$$\text{area} = 1.68 \text{ m}^2 \quad \text{and the diameter} = 1.46 \text{ m}$$

If the diameter is calculated from the velocity at the bottom of the column, the result is a diameter of 1·67 m so that if the velocity is not to exceed 1 m/s in any part of the column its diameter must be 1·67 m.

(e) The velocity based on the free area (tower area – downcomer area) must not exceed 0·75 m/s. The vapour rate in the bottom of the column is 2·17 m^3/s and, for a single-pass crossflow tray, the free area is approximately 88% of the tower area.

Then $\qquad\qquad\qquad A_t = 2.17/0.75 \times 0.88 = 3.28 \text{ m}^2$

and $\qquad\qquad\qquad D_t = 2.05 \text{ m}$

Problem 11.26

For a system that obeys Raoult's law show that the relative volatility α_{AB} is P_A^0/P_B^0, where P_A^0 and P_B^0 are the vapour pressures of the components A and B at the given temperature.

From the vapour pressure curves of benzene, toluene, ethyl benzene, and of *ortho-*, *meta-*, and *para*-xylenes, obtain a plot of the volatilities of each of the materials relative to *meta*-xylene over the range of 340–430 K.

Solution

Volatility of $A = P_A/x_A$ and the volatility of $B = P_B/x_B$ and the relative volatility α_{AB} is the ratio of these volatilities,

i.e.
$$\alpha_{AB} = \frac{P_A\, x_B}{x_A\, P_B}$$

For a system that obeys Raoult's law, $P = xP^0$.

$\therefore$
$$\alpha_{AB} = \frac{(x_A\, P_A^0)\, x_B}{x_A\, (x_B\, P_B^0)} = \frac{P_A^0}{P_B^0}$$

The vapour pressures of the compounds given in the problem are plotted in Fig. 11s

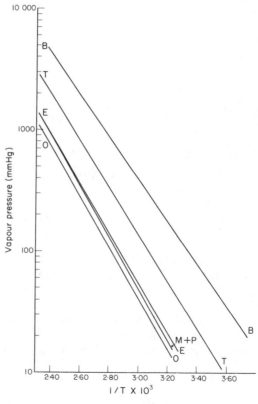

FIG. 11s

and are taken from Perry.[1] For convenience, the vapour pressures are plotted on a logarithmic scale against the reciprocal of the temperature (1/K) as a straight line then results. The relative volatilities may then be calculated in tabular form.

Temperature (K)	10^3/K	$m-x$	P^0 (mmHg)/$p-x$ $o-x$		E	B	T
340	2·94	65	64	56	72	490	180
360	2·78	141	140	118	155	940	360
380	2·63	292	290	240	310	1700	700
400	2·50	554	550	450	570	2900	1250
415	2·41	855	850	700	860	4200	1900
430	2·32	1315	1300	1050	1350	6000	2800

Temperature (K)	α_{pm}	α_{om}	α_{Em}	α_{Bm}	α_{Tm}
340	0·98	0·86	1·11	7·54	2·77
360	0·99	0·84	1·10	6·67	2·55
380	0·99	0·82	1·06	5·82	2·40
400	0·99	0·81	1·03	5·23	2·26
415	0·99	0·82	1·01	4·91	2·22
430	0·99	0·80	1·03	4·56	2·13

These data are plotted in Fig. 11t.

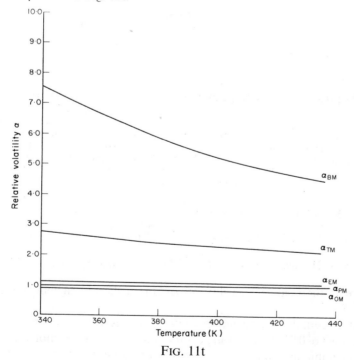

FIG. 11t

[1] PERRY, J. H.: *Chemical Engineers' Handbook*, 5th edn. (McGraw-Hill, 1973). Revised by R. PERRY and C. CHILTON.

Problem 11.27

A still has a liquor composition of *ortho*-xylene 10%, *meta*-xylene 65%, *para*-xylene 17%, benzene 4%, and ethyl benzene 4%. How many plates at total reflux are required to give a prod ict of 80% *meta*-xylene and 14% *para*-xylene? The data are given as mol per cent.

Solution

Fenske's equation may be used to find the number of plates at total reflux.

$$n + 1 = \frac{\log\left[(x_A/x_B)_d (x_B/x_A)_s\right]}{\log \alpha_{AB}} \qquad \text{(equation 11.42)}$$

In multicomponent distillation, A and B are the light and heavy key components respectively. In this problem, the only data given for both top and bottom products are for *meta*- and *para*-xylene and these will be used with the average relative volatility calculated in the previous problem. Thus:

$$x_A = 0.8, \quad x_B = 0.14, \quad x_{B_s} = 0.17, \quad x_{A_s} = 0.65$$

$$\alpha_{AB} = 1/0.99 = 1.0101$$

∴

$$n + 1 = \log\left[(0.8/0.14)(0.17/0.65)\right]/\log 1.0101$$

and

$$n = 39 \text{ plates}$$

Problem 11.28

The vapour pressure of n-pentane and of n-hexane is given in the table below.

Pressure (kN/m²)	1·3	2·6	5·3	8·0	13·3	26·6	53·2	101·3
(mmHg)	10	20	40	60	100	200	400	760
C_5H_{12} (Temp. in K)	223·1	233·0	244·0	251·0	260·6	275·1	291·7	309·3
C_6H_{14}	248·2	259·1	276·9	279·0	289·0	304·8	322·8	341·9

Equilibrium curve at atmospheric pressure:

$$x = 0.1 \quad 0.2 \quad 0.3 \quad 0.4 \quad 0.5 \quad 0.6 \quad 0.7 \quad 0.8 \quad 0.9$$

$$y = 0.21 \quad 0.41 \quad 0.54 \quad 0.66 \quad 0.745 \quad 0.82 \quad 0.875 \quad 0.925 \quad 0.975$$

(a) Determine the relative volatility of pentane to hexane at temperatures of 273, 293, and 313 K.

(b) A mixture containing 0·52 mol fraction pentane is to be distilled continuously to give a top product of 0·95 mol fraction pentane and a bottom of 0·1 mol fraction pentane. Determine the minimum number of plates (i.e. the number of plates at total reflux) by the graphical McCabe–Thiele method, and analytically by using the relative volatility method.

(c) Using the conditions as in (b), determine the liquid composition on the second plate from the top by Lewis's method if a reflux ratio of 2 is used.

(d) Using the conditions as in (b), determine by the McCabe–Thiele method the total number of plates required and the position of the feed.

It may be assumed that the feed is all liquid at the boiling-point.

Solution

The vapour pressure data and the equilibrium data are plotted in Figs. 11u and 11v.

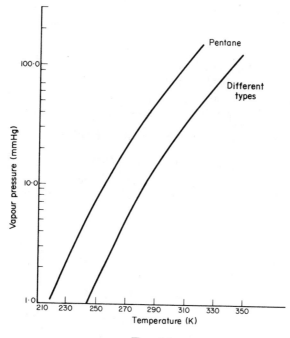

FIG. 11u

(a) Using
$$\alpha_{PH} = P_p^0/P_H^0$$

Temperature (K)	P_p^0	P_H^0	α_{PH}
273	24	6·0	4·0
293	55	16·0	3·44
313	116	36·5	3·18

Average = 3·54

(b) The McCabe–Thiele construction is shown in Fig. 11v where it is seen that 4 theoretical plates are required at total reflux.

Using Fenske's equation at total reflux:

$$n + 1 = \log\left[(0.95/0.05)(0.90/0.10)\right]/\log 3.54$$

and

$$\underline{n = 3.07}$$

(The discrepancy is caused by using an average value of α when α does in fact vary considerably.)

(c) From a mass balance it is found that for 100 kmol of feed and $R = 2$

$$D = 49.4, \quad W = 50.6, \quad L_n = 98.8, \quad V_n = 148.2$$

Then
$$y_n = \frac{L_n}{V_n}x_{n+1} + \frac{D}{V_n}x_d \qquad \text{(equation 11.19)}$$

and
$$y_n = 0.67x_{n+1} + 0.317$$

The vapour leaving the top plate has the composition of the distillate, i.e. $y_t = x_d = 0.95$. The liquid on the top plate is in equilibrium with this vapour and from the equilibrium curve has a composition $x_t = 0.845$.

The vapour rising to the top tray y_{t-1} is found from the operating line:

$$y_{t-1} = 0.67 \times 0.845 + 0.317 = 0.883$$

$$x_{t-1} = \text{from the equilibrium curve} = 0.707$$

$$y_{t-2} = 0.67 \times 0.707 + 0.317 = 0.790$$

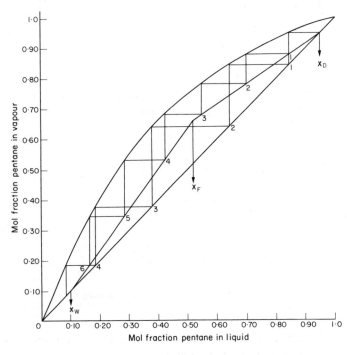

FIG. 11v

and
$$x_{t-2} = \underline{0.56}$$

(d) From Fig. 11v, 6 theoretical plates are required and the feed tray is the third from the top of the column.

Problem 11.29

The vapour pressures of n-pentane and n-hexane are given in the table below:

P^0 (kN/m²)	1·3	2·6	5·3	8·0	13·3	26·6	53·2	101·3
(mmHg)	10	20	40	60	100	200	400	760
C_5H_{12} Temp. in K	223·1	233·0	244·0	251·0	260·6	275·1	291·7	309·3
C_6H_{14}	248·2	259·1	276·9	279·0	289·0	304·8	322·8	341·9

Assuming that both Raoult's and Dalton's laws are obeyed:

(a) Plot the equilibrium curve for a total pressure of $13.3\,kN/m^2$.
(b) Determine the relative volatility of pentane to hexane as a function of liquid composition for a total pressure of $13.3\,kN/m^2$.
(c) Would the error caused by assuming the relative volatility constant at its mean value be considerable?
(d) Would it be more advantageous to distil this mixture at higher pressures?

Solution

(a) The following equations are used where A is n-pentane and B is n-hexane:

$$x_A = \frac{P - P_B^0}{P_A^0 - P_B^0} \qquad \text{(equation 11.5)}$$

$$y_A = P_A^0 x_A / P \qquad \text{(equation 11.4)}$$

At $P = 13.3\,kN/m^2$:

Temperature	P_A^0	P_B^0	x_A	y_A	$\alpha = P_A^0/P_B^0$
260·6	13·3	2·85	1·0	1·0	4·67
265	16·5	3·6	0·752	0·933	4·58
270	21·0	5·0	0·519	0·819	4·20
275	26·0	6·7	0·342	0·669	3·88
280	32·5	8·9	0·186	0·455	3·65
285	40·0	11·0	0·079	0·238	3·64
289	47·0	13·3	0	0	3·53

Mean $\alpha = \underline{4.02}$

These figures are plotted in Fig. 11w.
(b) The relative volatility is plotted as a function of liquid composition in Fig. 11w.

(c) If α is taken as 4·02, y_A may be calculated from:

$$y_A = \frac{\alpha x_A}{1 + (\alpha - 1) x_A} \qquad \text{(equation 11.8)}$$

Using equation 11.8, a new equilibrium curve may be calculated:

x_A	0	0·05	0·10	0·20	0·40	0·60	0·80	1·0
y_A	0	0·174	0·308	0·500	0·727	0·857	0·941	1·0

These points are shown in Fig. 11w where it is seen that little error is introduced by the use of this average value.

(d) If a higher pressure is contemplated, the method used in (a) is repeated. If $P = 100 \, \text{kN/m}^2$ is selected, the temperature range increases to 309–341 K and the new curve is drawn in Fig. 11w. Clearly, the higher pressure demands a larger number of plates for the same separation and is not desirable.

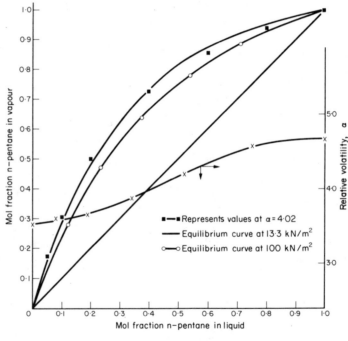

FIG. 11w

Problem 11.30

It is desired to separate a binary mixture by simple distillation. If the feed mixture has a composition of 0·5 mol fraction, calculate the fraction it is necessary to vaporise in order to obtain:

(a) a product of composition 0·75 mol fraction, when using a continuous process; and

(b) a product whose composition is not less than 0·75 mol fraction at any instant, when using a batch process.

If the product of the batch distillation is all collected in a single receiver, what is its mean composition?

Assume that the equilibrium curve is given by

$$y = 1\cdot2x + 0\cdot3$$

within the liquid composition range 0·3–0·8.

Solution

(a) If F = number of kmol of feed of composition x_f,

    L = kmol remaining in still with composition x,

    V = kmol of vapour formed with composition y, then

$$F = V + L \quad \text{and} \quad Fx_f = Vy + Lx$$

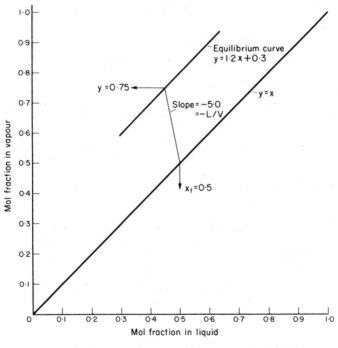

FIG. 11x

For 1 kmol of feed:

$$x_f = Vy + Lx \quad \text{and} \quad y = \frac{x_f}{V} - \frac{L}{V}x$$

This equation is a straight line of slope $- L/V$ which passes through the point (x_f, x_f), so that, knowing y, L/V may be found. This is illustrated in Fig. 11x where:

$$-L/V = -5 \cdot 0$$

as $$F = 1, \quad 1 = V + L$$

and $$V = 0 \cdot 167 \, \text{kmol/kmol of feed}$$

i.e. $$\underline{16 \cdot 7\% \text{ is vaporised}}$$

(b) For a batch process it can be shown that:

$$\frac{y - x}{y_0 - x_0} = \left(\frac{S}{S_0}\right)^{m-1} \qquad \text{(equation 11.13)}$$

where S is the number of kmols charged initially $= 100 \, \text{kmol}$ (say), S_0 is the number of kmols remaining, x is the initial still composition $= 0 \cdot 5$, y is the initial vapour composition $= (1 \cdot 2 \times 0 \cdot 5) + 0 \cdot 3 = 0 \cdot 90$, y_0 is the final vapour composition $= 0 \cdot 75$, x_0 is the final liquid composition and is found from

$$0 \cdot 75 = 1 \cdot 2 x_0 + 0 \cdot 3 \quad \text{or} \quad x_0 = 0 \cdot 375$$

and m is the slope of equilibrium curve $= 1 \cdot 2$.

$$\therefore \qquad \frac{0 \cdot 90 - 0 \cdot 50}{0 \cdot 75 - 0 \cdot 375} = 1 \cdot 07 = \left(\frac{100}{S_0}\right)^{0 \cdot 2}$$

and $$S_0 = 71 \cdot 3 \, \text{kmol/100 kmol feed}$$

$$\therefore \qquad \text{Amount vaporised} = \frac{100 - 71 \cdot 3}{100} \times 100 = \underline{28 \cdot 7\%}$$

The distillate composition may be found from a mass balance.

	Total kmol	kmol A	kmol B
Charge	100	50	50
Distillate	28·7	(50 − 26·7) = 23·3	(50 − 44·6) = 5·4
Residue	71·3	(0·374 × 71·3) = 26·7	(71·3 − 26·7) = 44·6

$$\therefore \qquad \text{Distillate composition} = (23 \cdot 3/28 \cdot 7) \times 100 = \underline{81 \cdot 2\%}$$

Problem 11.31

A liquor consisting of phenol and cresols with some xylenol is separated in a plate column. Given the following compositions, complete the material balance:

Mol %	Feed	Top	Bottom
C_6H_5OH	35	95·3	5·24
$o\text{-}C_7H_7OH$	15	4·55	
$m\text{-}C_7H_7OH$	30	0·15	
C_8H_9OH	20	—	
	100	100	

Also, calculate:

  (a) The composition on the second plate from the top.
  (b) The composition on the second plate from the bottom.

A reflux ratio of 4:1 is used.

Solution

The mass balance is completed as shown in Problem 11.24, where it was shown that:

$$x_{wo} = \underline{0·2017}, \quad x_{wm} = \underline{0·4472}, \quad x_{wx} = \underline{0·2987}$$

(a) For 100 kmol feed, from mass balances with $R = 4·0$, the following figures are obtained:

$$L_n = 132, \quad V_n = 165, \quad L_n = 232, \quad V_n = 165$$

The equations for the operating lines of each component are obtained from:

$$y_n = \frac{L_n}{V_n}x_{n+1} + \frac{D}{V_n}x_d \qquad \text{(equation 11.19)}$$

as:

$$P \quad y_{np} = 0·8x_{n+1} + 0·191$$

$$O \quad y_{no} = 0·8x_{n+1} + 0·009$$

$$M \quad y_{nm} = 0·8x_{n+1} + 0·0003$$

$$X \quad y_{nx} = 0·8x_{n+1}$$

The compositions on each plate may then be found calculating y from the operating line equations and x from:

$$x = \frac{y/\alpha}{\Sigma(y/\alpha)}$$

	α	$y_t = x_d$	y_t/α	$x_t = (y_t/\alpha)/0·81$	y_{t-1}	y_{t-1}/α	x_{t-1}
P	1·25	0·953	0·762	0·940	0·943	0·762	0·843
O	1·0	0·0455	0·046	0·057	0·055	0·055	0·061
M	0·63	0·0015	0·002	0·003	0·002	0·087	0·096
X	0·37	—	—	—	—	—	—
		1·000	$\Sigma = 0·81$	1·000	1·000	0·904	1·000

(b) In a similar way, the following operating lines may be derived for the bottom of the column as:

$$P \quad y_{mp} = 1{\cdot}406x_{n+1} - 0{\cdot}0212$$

$$O \quad y_{mo} = 1{\cdot}406x_{n+1} - 0{\cdot}0819$$

$$M \quad y_{mm} = 1{\cdot}406x_{n+1} - 0{\cdot}1816$$

$$X \quad y_{mx} = 1{\cdot}406x_{n+1} - 0{\cdot}1213$$

Then:

	α	x_s	αx_s	$y_s = \alpha x_s/0{\cdot}661$	x_1	αx_1	y_1	x_2
P	1·25	0·0524	0·066	0·100	0·086	0·108	0·148	0·121
O	1	0·2017	0·202	0·305	0·275	0·275	0·375	0·324
M	0·63	0·4472	0·282	0·427	0·433	0·273	0·373	0·394
X	0·37	0·2987	0·111	0·168	0·206	0·076	0·104	0·161
			$\Sigma = 0{\cdot}661$	1·000	1·000	0·732	1·000	1·000

Problem 11.32

A mixture of 60, 30, and 10 mol % benzene, toluene, and xylene respectively is separated by a plate column to give a top product containing at least 90 mol % benzene and negligible xylene, and a waste containing not more than 60 mol % toluene.

Using a reflux ratio of 4 and assuming that the feed is boiling liquid, determine the number of plates required in the column and the approximate position of the feed.

Take the relative volatility of benzene to toluene as 2·4 and of xylene to toluene as 0·45 and assume these values are constant throughout the column.

Solution

Assuming 100 kmol of feed, the mass balance may be completed to give:

$$D = 60, \quad W = 40\,\text{kmol}$$

$$x_{dt} = 0{\cdot}10, \quad x_{wb} = 0{\cdot}15, \quad x_{wx} = 0{\cdot}25$$

If $R = 4$ and the feed is at its boiling-point:

$$L_n = 240, \quad V_n = 300, \quad L_m = 340, \quad V_m = 300$$

and the top and bottom operating lines are found to be:

$$y_n = 0{\cdot}8x_{n+1} + 0{\cdot}2x_d$$

and

$$y_m = 1{\cdot}13x_{n+1} - 0{\cdot}133x_w$$

A plate-to-plate calculation may be carried out as follows:
(a) In the bottom of the column.

α		x_s	αx_s	y_s	x_1	αx_1	y_1	x_2	αx_2	y_2	x_3
2·4	B	0·15	0·360	0·336	0·314	0·754	0·549	0·503	1·207	0·724	0·657
1·0	T	0·60	0·600	0·559	0·564	0·564	0·411	0·432	0·432	0·258	0·298
0·45	X	0·25	0·113	0·105	0·122	0·055	0·040	0·065	0·029	0·018	0·045
			1·073	1·000	1·000	1·373	1·000	1·000	1·668	1·000	1·000

The composition on the third plate from the bottom corresponds most closely to the feed and above this tray the rectifying equations will be used.

	x_3	αx_3	y_3	x_4	αx_4	y_4
B	0·657	1·577	0·832	0·815	1·956	0·917
T	0·298	0·298	0·157	0·171	0·171	0·080
X	0·045	0·020	0·011	0·014	0·006	0·003
	1·000	1·895	1·000	1·000	2·133	1·000

As the vapour leaving the top plate will be totally condensed to give the product, 4 theoretical plates will be required to meet the given specification.

Problem 11.33

It is desired to concentrate a mixture of ethanol and water from 40 mol% to 70 mol% ethanol. A continuous fractionating column 1·2 m in diameter and having 10 plates is available. It is known that the optimum superficial vapour velocity in the column at atmospheric pressure is 1 m/s, giving an overall plate efficiency of 50%.

Assuming that the mixture is fed to the column as a boiling liquid and using a reflux ratio of twice the minimum value possible, determine the feed plate and the rate at which the mixture can be separated.

Equilibrium data:

Mol fraction alcohol in liquid:

0·1   0·2   0·3   0·4   0·5   0·6   0·7   0·8   0·89

Mol fraction alcohol in vapour:

0·43   0·526   0·577   0·615   0·655   0·70   0·754   0·82   0·89

Solution

The equilibrium data are plotted in Fig. 11y, where the operating line corresponding to the minimum reflux ratio is drawn from the point (x_d, x_d) through the intersection of the vertical q-line and the equilibrium curve to give an intercept of 0.505.

$$\therefore \qquad x_d/(R_m+1) = 0.505 \quad \text{and} \quad R_m = 0.386$$

The actual value of R is then $2 \times 0.386 = 0.772$ so that the top operating line may be constructed as shown.

This column contains the equivalent of (10×0.5), i.e. 5 theoretical plates, so that these may be stepped off from the point (x_d, x_d) to give the feed plate as the third from the top.

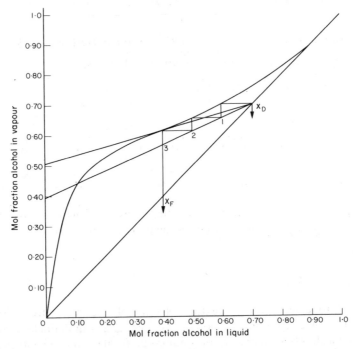

Fig. 11y

The problem as stated gives no bottom product composition, so that whilst all flow rates in the top of the column can be calculated, no information about the lower half can be derived. In the absence of these data, the feed rate cannot be determined, though the rate of distillate removal can as is shown below.

Mean molecular weight of top product $= (4.6 \times 0.7 + 18 \times 0.3)$

$$= 37.6 \,\text{kg/kmol}$$

If the top temperature is assumed to be 353 K, then:

$$\rho = (1/22.4)(273/353) = 0.0345 \,\text{kmol/m}^3$$

If the vapour velocity $= 1\,\text{m/s}$, the volumetric vapour flow at the top of the column $= (\pi/4)(1.2)^2 \times 1 = 1.13\,\text{m}^3/\text{s}$.

Hence $\qquad\qquad V_n = 1.13 \times 0.0345 = 0.039 \,\text{kmol/m}^3$

From the slope of the operating line,

$$L_n/V_n = 0.436$$

$$\therefore \qquad L_n = 0.436 \times 0.039 = 0.017\,\text{kmol/s}$$

As $R = 0.772$, $\qquad D = L_n/0.772 = 0.022\,\text{kmol/s}$

$$= \underline{\underline{0.828\,\text{kg/s distillate}}}$$

SECTION 12

ABSORPTION OF GASES

Problem 12.1

Some experiments are made on the absorption of carbon dioxide–air mixture in 2·5 normal caustic soda, using a 250 mm diameter tower packed to a height of 3 m with 19 mm Raschig rings.

In one experiment at atmospheric pressure, the results obtained were:

$$\text{gas rate } G' : 0\text{·}34 \, \text{kg/m}^2 \, \text{s}; \quad \text{liquid } L' : 3\text{·}94 \, \text{kg/m}^2 \, \text{s}$$

The carbon dioxide in the inlet gas is 315 parts per million and in the exit gas 31 parts per million.

What is the value of the overall gas transfer coefficient $K_G a$?

Solution

At the bottom of the tower:
$$y_1 = 315 \times 10^{-6}$$
$$G' = 0\text{·}34 \, \text{kg/m}^2 \, \text{s}$$
$$G_m = 0\text{·}34/29 = 0\text{·}0117 \, \text{kmol/m}^2 \, \text{s}$$

At the top of the tower:
$$y_2 = 31 \times 10^{-6}$$
$$x_2 = 0$$
$$L' = 3\text{·}94 \, \text{kg/m}^2 \, \text{s}$$

2·5N NaOH contains $2\text{·}5 \times 40 \, \text{g/l} = 100 \, \text{kg/m}^3$ NaOH.

Mean molecular weight of liquid

$$\frac{(100 \times 40) + (900 \times 18)}{1000} = 20\text{·}2 \, \text{kg/kmol}$$

$$\therefore \qquad L_m = 3\text{·}94/20\text{·}2 = 0\text{·}195 \, \text{kmol/m}^2 \, \text{s}$$

A mass balance over the tower gives (as $y \simeq Y$ for dilute gases)

$$G_m(y_1 - y_2) A = K_G \, aP(y - y_e)_{lm} \, ZA$$

It may be assumed that as the solution of NaOH is fairly concentrated, there will be negligible vapour pressure of CO_2 over the solution, i.e. all resistance to transfer lies in the gas phase.

$\therefore$ driving force at the top of the tower $= y_2 - 0 = 31 \times 10^{-6}$

and   at the bottom of the tower $= y_1 - 0 = 315 \times 10^{-6}$

Log mean driving force $(y - y_e)_{lm} = \dfrac{(315 - 31) \times 10^6}{\ln 315/31}$

$$= 122{\cdot}5 \times 10^{-6}$$

$\therefore$ $0{\cdot}0117(315 - 31) \, 10^{-6} = K_G a \times 101{\cdot}3 \times 122{\cdot}5 \times 10^{-6} \times 3$

from which $K_G a = \underline{8{\cdot}93 \times 10^{-5} \, \text{kmol/m}^3 \, \text{s} \, (\text{kN/m}^2)}$

$\checkmark$ **Problem 12.2**

An acetone–air mixture containing 0·015 mol fraction of acetone has the mol fraction reduced to 1% of this value by countercurrent absorption with water in a packed tower. The gas flow-rate G is 1 kg/m^2 s of air and the water entering is 1·6 kg/m^2 s. For this system, Henry's law holds and $y_e = 1{\cdot}75x$, where y_e is the mol fraction of acetone in the vapour in equilibrium with a mol fraction x in the liquid. How many overall transfer units are required?

Solution

As the system is dilute, mol fractions are approximately equal to mol ratios.

At the bottom of the tower: $y_1 = 0{\cdot}015$

$$G = 1{\cdot}0 \, \text{kg/m}^2 \, \text{s}$$

x_1 is unknown

At the top of the tower: $y_2 = 0{\cdot}000\,15$

$$x_2 = 0$$

$$L = 1{\cdot}6 \, \text{kg/m}^2 \, \text{s}$$

$\therefore$ $L_m = 1{\cdot}6/18 = 0{\cdot}0889 \, \text{kmol/m}^2 \, \text{s}$

            $G_m = 1{\cdot}0/29 = 0{\cdot}0345 \, \text{kmol/m}^2 \, \text{s}$

An overall mass balance gives:

$$G_m(y_1 - y_2) = L_m(x_1 - x_2)$$

or $0{\cdot}0345(0{\cdot}015 - 0{\cdot}000\,15) = 0{\cdot}0889(x_1 - 0)$

and $x_1 = 0{\cdot}005\,76$

$\therefore$ $y_{e_1} = 1{\cdot}75 \times 0{\cdot}005\,76 = 0{\cdot}0101$

The number of overall transfer units is defined by:

$$N_{OG} = \int_{y_2}^{y_1} \frac{dy}{y - y_e} = \frac{y_1 - y_2}{(y - y_e)_{lm}}$$

(from equations 12.67 and 12.77)

Top driving force $= y_2 - y_{e_2} = 0.000\,15$ since $x_2 = 0$.

Bottom driving force $= y_1 - y_{e_1} = 0.015 - 0.0101 = 0.0049$.

$\therefore$
$$(y - y_e)_{lm} = (0.0049 - 0.000\,15)/\ln(0.0049/0.000\,15) = 0.001\,36$$

$\therefore$
$$N_{OG} = (0.015 - 0.000\,15)/0.001\,36 = \underline{\underline{10.92}}$$

Also,
$$N_{OL} = N_{OG} \times mG_m/L_m$$

$$= 10.92 \times 1.75 \times 0.0345/0.0889 = \underline{\underline{7.42}}$$

✓ Problem 12.3

An oil containing 2·55 mol % of a hydrocarbon is stripped by running the oil down a column up which live steam is passed, so that 4 kmol of steam are used per 100 kmol of oil stripped. Determine the number of theoretical plates required to reduce the hydrocarbon content to 0·05 mol %, assuming that the oil is non-volatile. The vapour–liquid relation of the hydrocarbon in the oil is given by $y_e = 33x$, where y_e is the mol fraction in the vapour and x the mol fraction in the liquid. The temperature is maintained constant by internal heating, so that the steam does not condense in the tower.

Solution

If the steam does not condense,

$$L_m/G_m = 100/4 = 25$$

Inlet oil concentration $= 2.55$ mol %,

i.e.
$$x_2 = 0.0255 \quad \text{and} \quad X_2 = x_2/(1 - x_2) = 0.0262$$

Exit oil concentration $= 0.05$ mol %,

i.e.
$$x_1 = 0.0005$$

A mass balance between a plane in the tower, where the concentrations are x and y, and the bottom of the tower gives:

$$L_m(X - X_1) = G_m(Y - Y_1)$$

Now
$$Y_1 = 0$$

$\therefore$
$$Y = 25X - 25x_1$$

$$= 25X - 0.0125$$

This is the equation of the operating line and as the equilibrium data are $y_e = 33x$,

$$\therefore \quad \frac{Y}{1+Y} = \frac{33X}{1+X}$$

or

$$Y = \frac{33X}{1 - 32X}$$

Using these equilibrium data, the equilibrium and lines may be drawn as shown in Fig. 12a.

The number of theoretical plates is then found from a stepping-off procedure as employed for distillation as 8 plates.

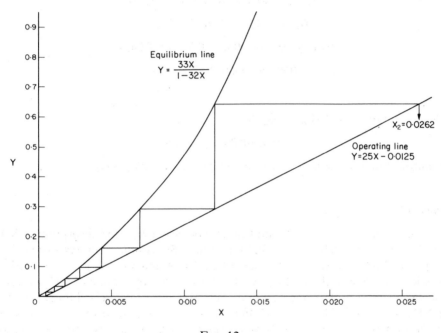

FIG. 12a

Problem 12.4

Gas from a petroleum distillation column has its concentration of H_2S reduced from 0·03 kmol H_2S/kmol of inert hydrocarbon gas to 1% of this value by scrubbing with a triethanolamine–water solvent in a countercurrent tower, operating at 300 K and at atmospheric pressure.

H_2S is soluble in such a solution and the equilibrium relation may be taken as $Y = 2X$, where Y is kmol of H_2S/kmol of inert gas and X is kmol of H_2S/kmol of solvent.

The solvent enters the tower free of H_2S and leaves containing 0·013 kmol of H_2S/kmol of solvent. If the flow of inert hydrocarbon gas is 0·015 kmol/m² s of tower cross-section and the gas-phase resistance controls the process, calculate:

(a) the height of the absorber necessary; and

(b) the number of transfer units required.

The overall coefficient for absorption $K''_G a$ may be taken as $0.04\,\text{kmol/s m}^3$ of tower volume (unit driving force in Y).

Solution

Driving force at top of column $= (Y_2 - Y_{2e}) = 0.0003$.

Driving force at bottom of column $= (Y_1 - Y_{1e}) = (0.03 - 0.026) = 0.004$.

Logarithmic mean driving force $= (0.004 - 0.0003)/\ln (0.004/0.0003) = 0.001\,43$.

From equation 12.70:

$$G_m(Y_1 - Y_2) A = K_G\, aP(Y - Y_e)_{\text{lm}} AZ$$

i.e.

$$G_m(Y_1 - Y_2) = K''_G\, a(Y - Y_e)_{\text{lm}}\, Z$$

$$\therefore \qquad 0.015(0.03 - 0.0003) = 0.04 \times 0.001\,43Z$$

$$Z = 0.000\,446/0.000\,057\,2 = 7.79\,\text{m}$$

$$= 7.8\,\text{m} \quad \text{(say)}$$

Height of transfer unit $\mathbf{H}_{OG} = G_m/K''_G\, a$

$$= (0.015/0.04) = 0.375\,\text{m}.$$

Number of transfer units $\mathbf{N}_{OG} = (7.79/0.375) = 20.8 = \underline{\underline{21}} \quad \text{(say)}.$

Problem 12.5

It is known that the overall liquid transfer coefficient $K_L a$ for absorption of SO_2 in water in a column is $0.003\,\text{kmol/s m}^3\,(\text{kmol/m}^3)$. By assuming an expression for the absorption of NH_3 in water at the same liquor rate and varying gas rates, derive an expression for the overall liquid film coefficient $K_L a$ for absorption of NH_3 in water in the same apparatus at the same water rate but varying gas rates. The diffusivities of SO_2 and NH_3 in air at $273\,K$ are 0.103 and $0.170\,\text{cm}^2/\text{s}$. SO_2 dissolves in water, so that Henry's constant is equal to $50\,(\text{kN/m}^2)/(\text{kmol/m}^3)$. All data are expressed for the same temperature.

Solution

From equation 12.18, $\dfrac{1}{K_L a} = \dfrac{1}{k_L a} + \dfrac{1}{Hk_G a} = \dfrac{1}{0.003} = 333.3$

For the absorption of a moderately soluble gas it is reasonable to assume that the liquid and gas phase resistances are equal, i.e.

$$\frac{1}{k_L a} = \frac{1}{Hk_G a} = \frac{333}{2} = 166.7$$

or $k_L a = Hk_G a = 0.006\,\text{kmol/s m}^3\,(\text{kmol/m}^3)$

∴ for SO_2, $k_G a = 0.006/H = 0.006/50 = 0.000\,12\,kmol/s\,m^3\,(kN/m^2)$

From equation 12.25, $k_G a$ is proportional to $(diffusivity)^{0.56}$ so that for NH_3:

$$k_G a = 0.000\,12\,(0.17/0.103)^{0.56} = 0.000\,16\,kmol/s\,m^3\,(kN/m^2)$$

(For a very soluble gas like NH_3, $k_G a \simeq K_G a$.)

For NH_3 the liquid film resistance will be small so that:

$$k_G a = K_G a = \underline{\underline{0.000\,16\,kmol/s\,m^3\,(kN/m^2)}}$$

Problem 12.6

A packed tower is used for absorbing sulphur dioxide from air by means of a N/2 caustic soda solution. At an air flow of $2\,kg/m^2\,s$, corresponding to a Reynolds number of 5160, the friction factor $R/\rho u^2$ is found to be 0.0200.

Calculate the mass transfer coefficient in kg $SO_2/s\,m^2\,(kN/m^2)$ under these conditions if the tower is at atmospheric pressure. At the temperature of absorption the following values may be taken:

Diffusion coefficient $SO_2 = 0.116\,cm^2/s$.
Viscosity of gas $= 0.018\,mN\,s/m^2$.
Density of gas stream $= 1.154\,kg/m^3$.

Solution

Work on wetted wall columns produced the following correlation:

$$\left(\frac{h_d}{u}\right)\left(\frac{P_{Bm}}{P}\right)\left(\frac{\mu}{\rho D}\right)^{0.56} = B' Re^{-0.17} = j_d \qquad \text{(equation 12.25)}$$

As discussed in Vol. 1:

$$j_D \simeq R/\rho u^2 \quad \text{(the friction factor)}$$

In this problem,

$$G' = 2.0\,kg/m^2\,s, \quad Re = 5160 \quad \text{and} \quad R/\rho u^2 = 0.020$$

$$D = 0.116 \times 10^{-4}\,m^2/s; \quad \mu = 1.8 \times 10^{-5}\,N\,s/m^2, \quad \text{and} \quad \rho = 1.154\,kg/m^3$$

Substituting these data:

$$\left(\frac{\mu}{\rho D}\right)^{0.56} = \left(\frac{1.8 \times 10^{-5}}{1.154 \times 0.116 \times 10^{-4}}\right)^{0.56} = 1.18$$

$$\left(\frac{h_d}{u}\right)\left(\frac{P_{Bm}}{P}\right) = 0.020/1.18 = 0.0169$$

$$G' = \rho u = 2.0\,kg/m^2\,s$$

$$u = 2.0/1.154 = 1.73\,m/s$$

$$h_d\,(P_{Bm}/P) = 0.0169 \times 1.73 = 0.0293$$

(At this point d can be found from $d = Re\,\mu/\rho u = 0.046$ m, which is the same order of size of wetted wall column that was originally used in the research work.)

Now
$$k_G = \left(\frac{h_d}{RT}\right)\left(\frac{P_{Bm}}{P}\right)$$

$R = 8.314\,\text{m}^3\,(\text{kN/m}^2)/\text{K kmol}$ and T will be taken as 298 K.

$$\therefore \qquad k_G = 0.0293/8.314 \times 298 = 1.18 \times 10^{-5}\,\text{kmol/m}^2\,\text{s}\,(\text{kN/m}^2)$$
$$= 7.56 \times 10^{-4}\,\text{kg SO}_2/\text{m}^2\,\text{s}\,(\text{kN/m}^2)$$

Problem 12.7

In an absorption tower, ammonia is being absorbed from air at atmospheric pressure by acetic acid. The flow rate of 2 kg/m² s in an experiment corresponds to a Reynolds number of 5100 and hence a friction factor $R/\rho u^2$ of 0.0199. At the temperature of absorption the viscosity of the gas stream μ is 0.018 mN/s m², the density ρ is 1.15 kg/m³, and the diffusion coefficient of ammonia in air D is 0.196 cm²/s.

Determine the mass transfer coefficient through the gas film in kg/m² s (kN/m²).

Solution

As in the previous problem:

$$\left(\frac{h_d}{u}\right)\left(\frac{P_{Bm}}{P}\right)\left(\frac{\mu}{\rho D}\right)^{0.56} = j_d \qquad\qquad \text{(equation 12.25)}$$

and
$$j_d \simeq R/\rho u^2$$

Substituting the given data gives:

$$(\mu/\rho D)^{0.56} = 0.88$$

$$\left(\frac{h_d}{u}\right)\left(\frac{P_{Bm}}{P}\right) = 0.0199/0.88 = 0.0226$$

$$u = G/\rho = 1.733\,\text{m/s}$$

$$\therefore \qquad k_G = \left(\frac{h_d}{RT}\right)\left(\frac{P_{Bm}}{P}\right) = \frac{0.0226 \times 1.733}{8.314 \times 298} = 1.58 \times 10^{-5}\,\text{kmol/m}^2\,\text{s}\,(\text{kN/m}^2)$$

$$k_G = 1.58 \times 10^{-5} \times 17 = 2.70 \times 10^{-4}\,\text{kg/m}^2\,\text{s}\,(\text{kN/m}^2)$$

Problem 12.8

Acetone is to be recovered from a 5% acetone–air mixture by scrubbing with water in a packed tower using countercurrent flow. The liquid rate is 0.85 kg/m² s and the gas rate is 0.5 kg/m² s.

The overall absorption coefficient K_G may be taken as $1.5 \times 10^{-4}\,\text{kmol/s m}^3\,(\text{kN/m}^2)$ partial pressure difference, and the gas film resistance controls the process.

What should be the height of the tower to remove 98% of the acetone? The equilibrium data for the mixture are:

Mol fraction acetone in gas:	0·0099	0·0196	0·0361	0·0400
Mol fraction acetone in liquid:	0·0076	0·0156	0·0306	0·0333

Solution

At the bottom of the tower: $y_1 = 0·05$

$$G = 0·95 \times 0·5 \, \text{kg/m}^2 \, \text{s}$$

$$G_m = 0·0164 \, \text{kmol/m}^2 \, \text{s}$$

At the top of the tower: $\quad y_2 = 0·02 \times 0·05 = 0·001$

$$L = 0·85 \, \text{kg/m}^2 \text{s}$$

$$L_m = 0·0472 \, \text{kmol/m}^2 \, \text{s}$$

The height and number of overall transfer units are defined as H_{OG} and N_{OG} by:

$$H_{OG} = G_m/K_G aP \quad \text{and} \quad N_{OG} = \int_{y_1}^{y_2} \frac{dy}{y - y_e}$$

(equations 12.80 and 12.77)

$$\therefore \qquad H_{OG} = 0·0164/(1·5 \times 10^{-4} \times 101·3) = 1·08 \, \text{m}$$

The equilibrium data in the statement of the problem represent a straight line of slope $m = 1·20$. As will be shown later in Problem 12.12, the equation for N_{OG} may be integrated directly when the equilibrium line is given by $y_e = mx$ to give:

$$N_{OG} = \frac{1}{(1 - mG_m/L_m)} \ln \left[\left(1 - \frac{mG_m}{L_m} \right) \frac{y_1}{y_2} + \frac{mG_m}{L_m} \right]$$

$$m(G_m/L_m) = 1·20(0·0164/0·0472) = 0·417$$

$$y_1/y_2 = 0·05/0·001 = 50$$

$$\therefore \qquad N_{OG} = \frac{1}{1 - 0·417} \ln \left[(1 - 0·417) \, 50 + 0·417 \right] = 5·80$$

$$\therefore \qquad \text{packed height} = N_{OG} \times H_{OG}$$

$$= 5·80 \times 1·08$$

$$= \underline{6·27 \, \text{m}}$$

Problem 12.9

Ammonia is to be removed from 10% ammonia–air mixture by countercurrent scrubbing with water in a packed tower at 293 K so that 99% of the ammonia is removed when working at a total pressure of 101·3 kN/m².

If the gas rate is $0.95\,\mathrm{kg/m^2\,s}$ of tower cross-section and the liquid rate is $0.65\,\mathrm{kg/m^2\,s}$, find the necessary height of the tower if the absorption coefficient $K_G a = 0.001\,\mathrm{kmol/m^3\,s}$ $(\mathrm{kN/m^2})$ partial pressure difference. The equilibrium data are:

kmol NH_3/kmol water:	0·021	0·031	0·042	0·053	0·079	0·106	0·159
Partial pressure NH_3:							
(mmHg)	12·0	18·2	24·9	31·7	50·0	69·6	114·0
$(\mathrm{kN/m^2})$	1·6	2·4	3·3	4·2	6·7	9·3	15·2

Solution

If the compositions of the gas are given as % by volume, at the bottom of the tower $y_1 = 0.10$ and $Y_1 = 0.10/(1-0.10) = 0.111$.

At the top of the tower, $y_2 = 0.001 \simeq Y_2$.

Mass flow rate of gas $= 0.95\,\mathrm{kg/m^2\,s}$.

% by weight of air $= [0.9 \times 29/(0.1 \times 17 + 0.9 \times 29)] \times 100 = 93.8\%$.

$\therefore$ mass flow rate of air $= 0.938 \times 0.95 = 0.891\,\mathrm{kg/m^2\,s}$

and $G_m = 0.891/29 = 0.0307\,\mathrm{kmol/m^2\,s}$

$$L_m = 0.65/18 = 0.036\,\mathrm{kmol/m^2\,s}$$

A mass balance between a plane in the tower where the compositions are X and Y and the top of the tower gives:

$$G_m(Y - Y_2) = L_m(X - X_2)$$

But $X_2 = 0$

$\therefore$ $0.0307(Y - 0.001) = 0.036X$ or $Y = 1.173X + 0.001$

This is the equation of the operating line in terms of mol ratios.

The given equilibrium data may be converted to the same basis since:

$$P_G = yP$$

$$= \frac{YP}{1 + Y}$$

and $Y = \dfrac{P_G}{P - P_G}$

kmol NH_3/kmol H_2O:	0·021	0·031	0·042	0·053	0·079	0·106	0·159
Partial pressure P_G (mm):	12	18·2	24·9	31·7	50·0	69·6	114·0
$P - P_G = 760 - P_G$:	748	741·8	735·1	728·3	710	690·4	646
$Y = P_G/(P - P_G)$:	0·016	0·0245	0·0339	0·0435	0·0704	0·101	0·176

These data are plotted in Fig. 12b.

From a mass balance over the column, the height Z is given by:

$$Z = \frac{G_m}{k_G aP} \int_{Y_2}^{Y_1} \frac{(1+Y)(1+Y_i)}{(Y-Y_i)} \, dY \qquad \text{(equation 12.51)}$$

Fig. 12b may be used to evaluate the integral.

Y	Y_i	$(1+Y)(1+Y_i)$	$\dfrac{(1+Y)(1+Y_i)}{(Y-Y_i)}$
0·111	0·089	1·21	55·0
0·10	0·078	1·185	53·8
0·08	0·059	1·14	54·3
0·06	0·042	1·11	61·4
0·04	0·027	1·067	82·0
0·02	0·013	1·035	148
0·01	0·006	1·016	254
0·005	0·0026	1·010	421
0·001	0	1·0	1000

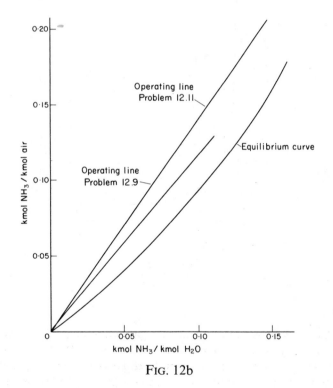

FIG. 12b

The area under the curve in Fig. 12c is found to be 12·6. For a very soluble gas $K_G a \simeq k_G a$ so that:

$$Z = \frac{0·0307}{0·001 \times 101·3} \times 12·6 = \underline{\underline{3·82\,\text{m}}}$$

[If the equilibrium line is assumed to be straight, then:

$$G_m(Y_2 - Y_1) = K_G aZ\Delta P_{lm}$$

Top driving force $= \Delta Y_2 = 0\cdot022$.
Bottom driving force $= \Delta Y_1 = 0\cdot001$.

$\therefore$ $$\Delta Y_{lm} = 0\cdot0068, \quad \Delta P_{lm} = 0\cdot688 \text{ kN/m}^2$$

and $$Z = \frac{0\cdot0307 \times 0\cdot11}{0\cdot001 \times 0\cdot688} = \underline{\underline{4\cdot91 \text{ m.}}}]$$

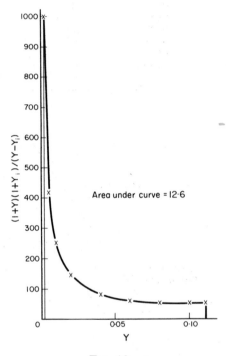

Area under curve = 12·6

FIG. 12c

Problem 12.10

Sulphur dioxide is recovered from a smelter gas containing $3\cdot5\%$ by volume of SO_2 by scrubbing with water in a countercurrent absorption tower. The gas is fed into the bottom of the tower, and in the exit gas from the top the SO_2 exerts a partial pressure of $1\cdot14$ kN/m². The water fed to the top of the tower is free from SO_2, and the exit liquor from the base contains $0\cdot001\,145$ kmol SO_2 per kmol water. The process takes place at 293 K at which the vapour pressure of water is $2\cdot3$ kN/m². The water rate is $0\cdot43$ kmol/s.

If the area of the tower is $1\cdot35$ m² and the overall coefficient of absorption for these conditions $K''_L a$ is $0\cdot19$ kmol SO_2/s m³ (kmol of SO_2 per kmol H_2O), calculate the necessary height of the column.

The equilibrium data for SO_2 and water at 293 K are:

kmol SO_2 per 1000 kmol H_2O:	0·056	0·14	0·28	0·42	0·56	0·84	1·405
kmol SO_2 per 1000 kmol inert gas:	0·7	1·6	4·3	7·9	11·6	19·4	36·3

Solution

At the top of the column: $\qquad P_{SO_2} = 1·14 \,\text{kN/m}^2$

i.e. $\qquad 1·14 = 101·3 y_2$ and $y_2 = 0·0113 \simeq Y_2$

At the bottom of the column: $y_1 = 0·035$, i.e. $Y_1 = 0·036$

$$X_1 = 0·001\,145$$

$$L_m = 0·43 \,\text{kmol/s}$$

The quantity of SO_2 absorbed $= 0·43(0·001\,145 - 0)$

i.e. $\qquad N_A = 4·94 \times 10^{-4} \,\text{kmol SO}_2/\text{s}$

Now $\qquad N_A = K_L'' a (X_e - X)_{lm}$

The log mean driving force in terms of the liquid phase must now be calculated. The values of X_e corresponding to the gas composition Y can be found from the equilibrium data given (but not plotted here) as:

When $\qquad Y_2 = 0·0113, \quad X_{e_2} = 0·54 \times 10^{-3}$

$\qquad Y_1 = 0·036, \quad X_{e_1} = 1·41 \times 10^{-3}$

$\therefore \qquad (X_{e_1} - X_1) = (1·41 - 1·145) 10^{-3} = 0·265 \times 10^{-3} \,\text{kmol SO}_2/\text{kmol H}_2\text{O}$

$(X_{e_2} - X_2) = 0·5 \times 10^{-3} \,\text{kmol SO}_2/\text{kmol H}_2\text{O}$

$\therefore \qquad (X_e - X)_{lm} = \dfrac{(0·54 - 0·265) 10^{-3}}{\ln (0·54/0·265)} = 3·86 \times 10^{-4} \,\text{kmol SO}_2/\text{kmol H}_2\text{O}$

$\therefore \qquad 4·94 \times 10^{-4} = 0·19 V \times 3·86 \times 10^{-4} \quad (V = \text{packed volume, m}^3)$

and $\qquad V = 6·74 \,\text{m}^3$

$\therefore \qquad$ packed height $= 6·74/1·35 = \underline{\underline{5·0 \,\text{m}}}$

Problem 12.11

Ammonia is removed from a 10% ammonia–air mixture by scrubbing with water in a packed tower, so that 99·9% of the ammonia is removed. What is the required height of the tower?

Entering gas: $1·2 \,\text{kg/m}^2 \,\text{s}$.

Water rate: $0·94 \,\text{kg/m}^2$.

$K_G a = 0·0008 \,\text{kmol/s m}^3 \,(\text{kN/m}^2)$.

Solution

The molecular weights of ammonia and air are 17 and 29 kg/kmol respectively. The data in wt. % must be converted to mol ratios as the inlet gas concentration is high.

$$\therefore \quad 0 \cdot 10 = \frac{17 y_1}{17 y_1 + 29 (1 - y_1)}$$

and

$$y_1 = 0 \cdot 159$$

$$Y_1 = \frac{0 \cdot 159}{1 - 0 \cdot 159} = 0 \cdot 189$$

$$Y_2 \simeq y_2 = 0 \cdot 000 \, 159$$

Entering gas rate $= 1 \cdot 2 \, \text{kg/m}^2 \, \text{s}$.
Entering ammonia $= 0 \cdot 12 \, \text{kg/m}^2 \, \text{s}$.
Entering air $= 1 \cdot 08 \, \text{kg/m}^2 \, \text{s}$.

$$\therefore \quad G_m = 0 \cdot 0372 \, \text{kmol/m}^2 \, \text{s}$$

$$L_m = 0 \cdot 94/18 = 0 \cdot 0522 \, \text{kmol/m}^2 \, \text{s}$$

$$X_2 = 0 \quad \text{(ammonia free)}$$

The equation of the operating line is found from a mass balance between a plane where the compositions are X and Y and the top of the tower as:

$$0 \cdot 0372 (Y - 0 \cdot 000 \, 159) = 0 \cdot 0522 X$$

or

$$Y = 1 \cdot 4 X + 0 \cdot 000 \, 159$$

This equilibrium line is plotted on Fig. 12b (Problem 12.9). As in Problem 12.9, the integral in the following equation may be obtained graphically from Fig. 12d as 40·55 using the data below.

$$Z = \frac{G_m}{k_G \, aP} \int_{Y_2}^{Y_1} \frac{(1 + Y)(1 + Y_i) \, dY}{(Y - Y_i)}$$

Y	Y_i	$(Y - Y_i)$	$(1 + Y)(1 + Y_i)$	$\dfrac{(1 + Y)(1 + Y_i)}{(Y - Y_i)}$
0·20	0·152	0·048	3·18	66·3
0·19	0·138	0·052	1·35	26·0
0·15	0·102	0·048	1·27	26·4
0·10	0·063	0·037	1·17	31·6
0·05	0·028	0·022	1·08	49·1
0·04	0·022	0·018	1·06	58·8
0·03	0·016	0·014	1·05	74·7
0·02	0·011	0·009	1·03	114·6
0·01	0·005	0·005	1·015	203·0
0·000 15	0·000	0·000 15	1·000 15	6670·0

$K_G a \simeq k_G a$ for a very soluble gas so that:

$$Z = \frac{0 \cdot 0372 \times 40 \cdot 55}{0 \cdot 0008 \times 101 \cdot 3} = \underline{\underline{18 \cdot 6 \, \text{m}}}$$

[It is interesting to note that if $Y = 0.01$ rather than $0.000\,15$, the integral has a value of 8.25 and Z is equal to 3.8 m. Thus 14.8 m of packing is required to remove the last traces of ammonia.]

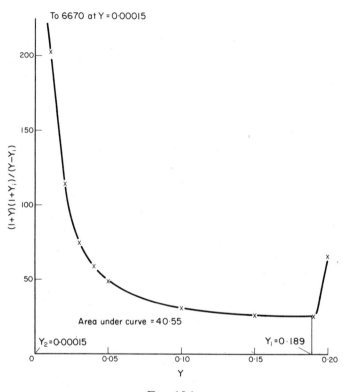

To 6670 at Y = 0·00015

Area under curve = 40·55

$Y_2 = 0.00015$

$Y_1 = 0.189$

Y

Fɪɢ. 12d

Problem 12.12

A soluble gas is absorbed from a dilute gas–air mixture by countercurrent scrubbing with a solvent in a packed tower. If the liquor led to the top of the tower contains no solute, show that the number of transfer units required is given by:

$$N = \left[1 \bigg/ \left(1 - \frac{mG_m}{L_m} \right) \right] \ln \left[\left(1 - \frac{mG_m}{L_m} \right) \frac{y_1}{y_2} + \frac{mG_m}{L_m} \right]$$

where G_m and L_m are the flow rates of the gas and liquid in kmols per s per m² of tower area and y_1 and y_2 the mol fraction of the gas at the inlet and outlet of the column. The equilibrium relation between the gas and liquid is represented by a straight line with the equation $y_e = mx$, where y_e is the mol fraction in the gas in equilibrium with x mol fraction in the liquid.

In a given process it is desired to recover 90% of the solute by using 50% more liquid than the minimum necessary. If the HTU of the proposed tower is 0.6 m, what height of packing will be required?

Solution

By definition

$$N_{OG} = \int_{y_2}^{y_1} \frac{dy}{y - y_e}$$

A mass balance between the top and some plane in the tower where the mol fractions are x and y gives:

$$G_m(y - y_2) = L_m(x - x_2)$$

If the inlet liquid is solute free, $x_2 = 0$ and:

$$x = \frac{G_m}{L_m}(y - y_2)$$

If the equilibrium data are represented by:

$$y_e = mx$$

substituting for $y_e = m(G_m/L_m)(y - y_2)$ gives:

$$N_{OG} = \int_{y_2}^{y_1} \frac{dy}{y - \dfrac{mG_m}{L_m}(y - y_2)}$$

$$= \int_{y_2}^{y_1} \frac{dy}{y\left(1 - \dfrac{mG_m}{L_m}\right) + \dfrac{mG_m}{L_m}y_2}$$

$$= \left[1 \middle/ \left(1 - \frac{mG_m}{L_m}\right)\right] \ln\left[\left(1 - \frac{mG_m}{L_m}\right)\frac{y_1}{y_2} + \frac{mG_m}{L_m}\right]$$

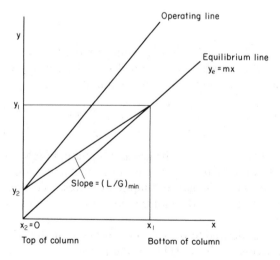

FIG. 12e

Referring to Fig. 12e:

$$\left(\frac{L}{G}\right)_{min} = \frac{y_1 - y_2}{x_1} = \frac{y_1 - y_2}{y_1/m} = m\left(1 - \frac{y_2}{y_1}\right)$$

$$= m\left(1 - \frac{0\cdot1y_1}{y_1}\right)$$

$$= 0\cdot9\,m$$

If $1\cdot5(L/G)_{min}$ is actually employed,

$$L_m/G_m = 1\cdot5 \times 0\cdot9\,m = 1\cdot35\,m$$

$$\therefore \qquad \frac{mG_m}{L_m} = \frac{m}{1\cdot35\,m} = 0\cdot74$$

$$y_1/y_2 = 10$$

$$\therefore \qquad N_{OG} = \frac{1}{0\cdot26}\ln\left[(0\cdot26 \times 10) + 0\cdot74\right] = 4\cdot64$$

$$H_{OG} = 0\cdot6\,m$$

$$\text{height of packing} = 0\cdot6 \times 4\cdot64 = \underline{\underline{2\cdot78\,m}}$$

Problem 12.13

A paraffin hydrocarbon of molecular weight 114 kg/kmol at a temperature of 373 K is to be separated from a mixture with a non-volatile organic compound of molecular weight 135 kg/kmol by stripping with steam. The liquor contains 8% of the paraffin by weight and this is to be reduced to 0·08% using an upward flow of steam saturated at 373 K. If three times the minimum amount of steam is used, how many theoretical stages will be required?

The vapour pressure of the paraffin at 373 K is 53 kN/m² and the process takes place at atmospheric pressure. It may be assumed that the system obeys Raoult's law.

Solution

If Raoult's law applies:

$$\text{Partial pressure} = x\,(\text{vapour pressure})$$

$$P_A = xP_A^0$$

Now

$$y = P_A/P$$

$$\therefore \qquad y_e = x(P_A^0/P)$$

$$= (53/101\cdot3)\,x = 0\cdot523x$$

In terms of mol ratios:

$$\frac{Y_e}{1 + Y_e} = \frac{0\cdot523X}{1 + X}$$

Thus the equilibrium curve can be derived.

X	$X/(1 + X)$	$Y_e/(1 + Y_e)$	Y_e
0	0	0	0
0·02	0·0196	0·0103	0·0104
0·04	0·0385	0·020	0·0204
0·06	0·0566	0·0296	0·0305
0·08	0·0741	0·0387	0·0403
0·10	0·0909	0·0475	0·0499
0·12	0·107	0·0560	0·059

This curve is plotted in Fig. 12f.
As the inlet contains 8% by weight of paraffin:

$$X_2 = (8/114)/(92/135) = 0{\cdot}103$$

and
$$X_1 = 0{\cdot}001\ 03$$

$$Y_1 = 0$$

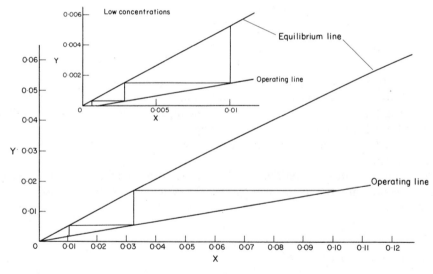

FIG. 12f

The minimum amount of steam occurs when the exit streams are in equilibrium.
When $X_2 = 0{\cdot}103$, $\qquad\qquad Y_{e_2} = 0{\cdot}0513$

From an overall mass balance:

$$L_{min}(0{\cdot}103 - 0{\cdot}001\ 03) = G_{min}(0{\cdot}0513 - 0)$$

and
$$(L/G)_{min} = 0{\cdot}503$$

$\therefore$
$$(L/G)\ \text{actual} = 0{\cdot}167$$

Thus the operating line passing through the point (0·001 03, 0) with a slope = 0·167 may be drawn in Fig. 12f and the number of theoretical stages drawn and found to be 4.

This problem may also be solved analytically by the use of the absorption factor method. This is illustrated later in Problem 12.16.

Problem 12.14

Benzene is to be absorbed from coal gas by means of a wash oil. The inlet gas contains 3% by volume of benzene, and the exit gas should not contain more than 0·02% benzene by volume. The suggested oil circulation rate is 490 kg oil per 100 m³ of inlet gas measured at STP. The wash oil enters the tower solute-free. If the overall height of a transfer unit based on the gas phase is 1·4 m, determine the minimum height of the tower which is required to carry out the absorption. Use the following equilibrium data:

Benzene in oil (% by weight):

	0·05	0·10	0·50	1·0	2·0	3·0

Equilibrium partial pressure of benzene in gas:

kN/m²:	0·013	0·033	0·020	0·53	1·33	3·33
mmHg:	0·1	0·25	1·5	4·0	10·0	25·0

Solution

At the top and bottom of the tower respectively:

$$y_2 = 0.0002, \quad x_2 = 0$$

and $\qquad\qquad y_1 = 0.03, \qquad x_1 = $ exit oil composition

Take 100 m³ of inlet gas at STP as the basis of calculation.
Volume of benzene at inlet = 0·03 × 100 = 3·0 m³ (97 m³ of gas).
Volume of benzene at exit = 0·0002 × 97 = 0·0194 m³.
At STP, kg molecular volume = 22·4 m³.

$\therefore\qquad\qquad$ kmol of gas = 97/22·4 = 4·33 kmol

Volume of benzene absorbed = 3·0 − 0·0194 = 2·9806 m³.
Density of benzene at STP = 78/22·4 = 3·482 kg/m³.

$\therefore\qquad\qquad$ weight of benzene absorbed = 2·9806 × 3·482

$$= 10.38 \text{ kg}$$

As oil rate = 490 kg/100 m³ of gas,

wt. % of benzene at exit = (10·38 × 100)/490 = 2·12%.

Now $\qquad\qquad\qquad Y_1 = 0.03/(1 - 0.03) = 0.031$

$$Y_2 \simeq y_2 = 0.0002$$

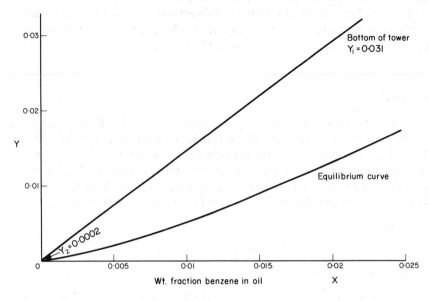

FIG. 12g

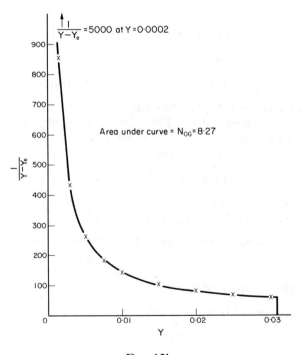

FIG. 12h

Thus the operating line can be plotted as shown in Fig. 12g. The equilibrium data given in the problem and converted to the appropriate units are as shown below and plotted on Fig. 12g.

Wt. % benzene	Wt. fraction	Equilibrium partial pressure (kN/m^2)	$Y_e = P^0/P$
0	0	0	0
0·05	0·0005	0·013	0·000 13
0·10	0·001	0·033	0·000 33
0·50	0·005	0·20	0·001 97
1·0	0·01	0·53	0·005 23
2·0	0·02	1·33	0·013 13
3·0	0·03	3·33	0·3287

$$N_{OG} = \int_{Y_1}^{Y_2} \frac{dY}{Y - Y_e}$$

The value of this integral may be evaluated from the operating and equilibrium lines by graphical means for values of Y between 0·0002 and 0·031.

Y	Y_e	$(Y - Y_e)$	$1/(Y - Y_e)$
0·0002	0	0·002	5000
0·0015	0·000 33	0·001 17	855
0·003	0·007	0·0023	435
0·005	0·0012	0·0038	263
0·0075	0·0021	0·0054	185
0·010	0·0031	0·0069	145
0·015	0·0055	0·0095	105
0·020	0·008	0·012	83
0·025	0·0106	0·0144	69
0·030	0·0137	0·0163	61

From Fig. 12h, the area under the curve $= 8·27 = N_{OG}$

$$\therefore \quad \text{height} = N_{OG} \times H_{OG}$$

$$= 8·27 \times 1·4$$

$$= \underline{11·6\,m}$$

Problem 12.15

Ammonia is to be recovered from a 5% by volume ammonia–air mixture by scrubbing with water in a packed tower. The gas rate is $1·25\,m^3/s\,m^2$ measured at NTP and the liquid rate is $1·95\,kg/m^2\,s$. The temperature of the inlet gas is 295 K and of the inlet water 293 K. The mass transfer coefficient is $K_G a = 0·113\,kmol/m^3\,s$ (mol ratio difference) and the total pressure is $101·3\,kN/m^2$. Find the height of the tower to remove 95% of the ammonia. The equilibrium data and the heats of solution are as follows:

Mol fraction in liquid:

$$0.005 \quad 0.01 \quad 0.015 \quad 0.02 \quad 0.03$$

Integral heat of solution (MJ/kmol of solution):

$$181 \quad 363 \quad 544 \quad 723 \quad 1084$$

Equilibrium partial pressure:

mmHg at 293 K: 3·0 (0·4)   5·8 (0·77)   8·7 (1·16)   11·6 (1·55)   17·5 (2·33)

(kN/m²) at 298 K: 3·6 (0·48)   7·3 (0·97)   10·7 (1·43)   14·4 (1·92)   22·0 (2·93)

(kN/m²) at 303 K: 4·6 (0·61)   9·6 (1·28)   13·7 (1·33)   18·5 (2·47)   29·0 (3·86)

Assume adiabatic conditions and neglect the heat transfer between phases.

Solution

The data provided are presented in Fig. 12i.
The entering gas rate $= 12.5 \, \text{m}^3/\text{m}^2 \, \text{s}$ at NTP.
Density at NTP $= 1/22.4 = 0.0446 \, \text{kmol/m}^3$.

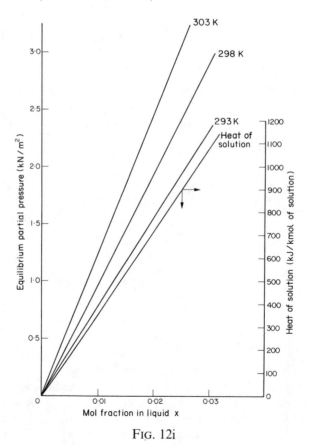

FIG. 12i

At the bottom of the tower, $y_1 = 0.05$,

$\therefore$ $\qquad Y_1 = 0.05/0.95 = 0.0526$

$\qquad G_m = 0.95 \times 1.25 \times 0.0446 = 0.053 \, \text{kmol/m}^2 \, \text{s}$

$\qquad Y_2 = 0.05 \times 0.0526 = 0.00263$

$\qquad L_m = 1.95/18 = 0.108 \, \text{kmol/m}^2 \, \text{s}$

$\qquad X_2 = 0$

An overall mass balance gives:

$$0.108(X_1 - 0) = 0.053(0.0526 - 0.00263)$$

from which $\qquad\qquad X_1 = 0.0245$

In this problem, the temperature varies throughout the column and the tower will be divided into increments so that by heat and mass balances the terminal conditions over each section may be found. Knowing the compositions and the temperature, the data given may be used in conjunction with the mass coefficient to calculate the height of the chosen increment. Adiabatic conditions will be assumed and, as the sensible heat change of the gas is small, the heat of solution will only be used to raise the temperature of the liquid. The gas temperature will therefore remain constant at 295 K.

Consider conditions at the top of the tower:

$$X_2 = 0, \quad T_1 = 293 \, \text{K}, \quad T_g = 295 \, \text{K}, \quad Y_2 = 0.00263$$

Choosing an increment such that the exit liquor stream and inlet gas streams have compositions $X = 0.005$ and Y, a mass balance taking 1 m^2 as a basis gives:

$$L_m(X - 0) = G_m(Y - Y_2)$$

i.e. $\qquad\qquad 0.108 \times 0.005 = 0.053(Y - 0.00263)$

from which $\qquad\qquad Y = 0.0128$

NH$_3$ absorbed in the section $= 0.108 \times 0.005 = 5.4 \times 10^{-4} \, \text{kmol/s}$.
Heat of solution when $X = 0.005 = 181 \, \text{kJ/kmol}$ of solution.

$\therefore$ $\qquad$ heat liberated $= 181 \times 0.005 = 19.55 \, \text{kW}$

$\therefore$ $\qquad$ temperature rise $= 19.55/(0.108 \times 18 \times 4.18)$

$$= 2.4 \, \text{degK}$$

and the liquid exit temperature $= 295.4 \, \text{K}$

At 295.4 K when $X = 0.005$,

$$P_e = 0.44 \, \text{kN/m}^2, \quad \therefore \, Y_e = 0.00434$$

At the top of the section, $Y - Y_e = 0.00263 - 0 = 0.00263$.
At the bottom of the section, $Y - Y_e = 0.0128 - 0.00434 = 0.00846$.

$\therefore$ $\qquad (Y - Y_e)_{\text{lm}} = (0.00846 - 0.00263)/\ln(0.00846/0.00263)$

$$= 0.00499$$

$\therefore$ $5.4 \times 10^{-4} = 0.113 \times 1 \times H \times 0.004\,99$ (since $A = 1\ m^2$)

and $H = 0.958\ m$

In an exactly similar way further increments may be taken and the heights of each found. A summary of these calculations is shown in the table.

Increment No.	Inlet X	Inlet Y	Outlet X	Outlet Y	Outlet liquid temp. (K)	Outlet P_e (kN/m²)	$(Y-Y_e)$ top	$(Y-Y_e)$ bottom	$(Y-Y_e)_{lm}$	Quantity absorbed (kmol/s)	Height of section (m)
1	0	0.002 63	0.005	0.0128	295.4	0.44	0.002 63	0.008 46	0.004 99	5.4×10^{-4}	0.958
2	0.005	0.0128	0.010	0.023	297.8	0.95	0.008 46	0.0136	0.010 83	5.4×10^{-4}	0.441
3	0.010	0.023	0.015	0.0332	300.2	1.62	0.0136	0.0172	0.0153	5.4×10^{-4}	0.328
4	0.015	0.0332	0.020	0.0434	302.6	2.40	0.0172	0.0197	0.0185	5.4×10^{-4}	0.258
5	0.020	0.0434	0.0245	0.0526	304.8	3.30	0.0197	0.020	0.0198	5.4×10^{-4}	0.241

$= 2.23\ m$

Thus the required height of packing $= \underline{\underline{2.23\ m}}$.

Problem 12.16

A thirty-plate bubble-cap column is to be used to remove n-pentane from solvent oil by means of steam stripping. The inlet oil contains 6 kmol of n-pentane per 100 kmol of pure oil and it is desired to reduce the solute content to 0·1 kmol per 100 kmol of solvent. Assuming isothermal operation and an overall plate efficiency of 30%, find the specific steam consumption, i.e. kmol of steam required per kmol of solvent oil treated, and the ratio of the specific and minimum steam consumptions. How many plates would be required if this ratio were 2·0?

The equilibrium relation for the system may be taken as $Y_e = 3.0X$, where Y_e and X are expressed in mol ratios of pentane in the gas and liquid phases respectively.

Solution

Number of theoretical plates $= 30 \times 0.3 = 9$.

At the bottom of the tower: kmol of steam $= G_m$

mol ratio of pentane in steam $= Y_1$

mol ratio of pentane in oil $= X_1 = 0.001$

At the top of the tower: exit steam composition $= Y_2$

inlet oil composition $= X_2 = 0.06$

kmol of oil $= L_m$

The minimum steam consumption occurs when the exit steam stream is in equilibrium with the inlet oil, i.e.

when $Y_{e_2} = 0.06 \times 3 = 0.18$

then $L_{min}(X_2 - X_1) = G_{min}(Y_2 - Y_1)$

If $Y_1 = 0$, i.e. no pentane in inlet steam,

$$L_{min}(0{\cdot}06 - 0{\cdot}001) = G_{min} \times 0{\cdot}18$$

$$\therefore \qquad (G/L)_{min} = (0{\cdot}06 - 0{\cdot}001)/0{\cdot}18 = 0{\cdot}328$$

The operating line can be fixed by trial and error as it passes through the point $(0{\cdot}001, 0)$ and 9 theoretical plates are required for the separation. Thus it is a matter of selecting the operating line which, with 9 steps, will give $X_2 = 0{\cdot}001$ when $X_1 = 0{\cdot}06$. This is tedious but possible, and the problem may be solved analytically since the equilibrium line is straight.

Use may be made of the absorption factor method where it is shown in equation 12.96 that

$$\frac{Y_1 - Y_2}{Y_1 - mX_2} = \frac{\mathscr{A}^{N+1} - \mathscr{A}}{\mathscr{A}^{N+1} - 1} \qquad \text{(equation 12.96)}$$

where $\mathscr{A}$ is the absorption factor $= L_m/mG_m$ and N is the number of theoretical plates.

The corresponding expression for a stripping operation is:

$$\frac{X_2 - X_1}{X_2 - Y_1/m} = \frac{(1/\mathscr{A})^{N+1} - (1/\mathscr{A})}{(1/\mathscr{A})^{N+1} - 1}$$

In this problem, $N = 9$, $X_2 = 0{\cdot}06$, $X_1 = 0{\cdot}001$, and $Y_1 = 0$

$$\therefore \qquad \frac{0{\cdot}06 - 0{\cdot}001}{0{\cdot}06} = 0{\cdot}983 = \frac{(1/\mathscr{A})^{10} - (1/\mathscr{A})}{(1/\mathscr{A})^{10} - 1}$$

from which by trial and error $\qquad 1/\mathscr{A} = 1{\cdot}37$

i.e. $\qquad \dfrac{mG_m}{L_m} = 1{\cdot}37, \quad \dfrac{G_m}{L_m} = \dfrac{1{\cdot}37}{3} = 0{\cdot}457$

$$\therefore \qquad \frac{\text{actual } G_m/L_m}{\text{minimum } G_m/L_m} = \frac{0{\cdot}457}{0{\cdot}328} = \underline{1{\cdot}39}$$

If $(\text{actual } G_m/L_m)/(\text{min } G_m/L_m) = 2$, actual $G_m/L_m = 0{\cdot}656$.

$$\therefore \qquad 1/\mathscr{A} = mG_m/L_m = 1{\cdot}968$$

$$\therefore \qquad 0{\cdot}983 = \frac{(1{\cdot}968)^{N+1} - 1{\cdot}968}{(1{\cdot}968)^{N+1} - 1}$$

from which $N = 4{\cdot}9$ and the actual number of plates $= 4{\cdot}9/0{\cdot}3 = 16{\cdot}3$ (say 17).

Problem 12.17

A mixture of ammonia and air is scrubbed in a plate column with fresh water.

If the ammonia concentration is reduced from 5% to 0·01%, and the water and air rates are respectively 0·65 and 0·40 kg/m² s, how many theoretical plates are required? The equilibrium relationship can be written $Y = X$, where X is the mol ratio in the liquid phase.

Solution

Assuming that the compositions are given as volume %;

At the bottom of the tower: $y_1 = 0.05$, $Y_1 = \dfrac{0.05}{1 - 0.05} = 0.0526$

At the top of the tower: $y_2 = 0.0001 = Y_2$

$$L_m = 0.65/18 = 0.036 \,\text{kmol/m}^2\,\text{s}$$

$$G_m = 0.40/29 = 0.0138 \,\text{kmol/m}^2\,\text{s}$$

A mass balance gives the equation of the operating line as:

$$0.0138(Y - 0.0001) = 0.036(X - 0)$$

or $Y = 2.61 + 0.0001$

The operating line and equilibrium line are then drawn and from Fig. 12j, 6 theoretical stages are required.

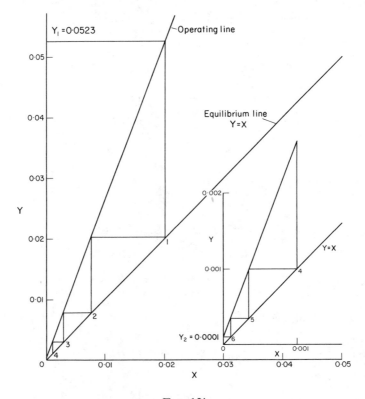

FIG. 12j

LIQUID–LIQUID SYSTEMS

Problem 13.1

Tests are made on the extraction of acetic acid from a dilute aqueous solution by means of a ketone in a small spray tower of diameter 46 mm and effective height 1·09 m, the aqueous phase being run into the top of the tower. The ketone enters free from acid at a rate of 0·0014 m³/s m² and leaves with an acid concentration of 0·38 kmol/m³. The concentration in the aqueous phase falls from 1·19 to 0·82 kmol/m³.

Calculate the overall extraction coefficient based on the concentrations in the ketone phase and the height of the corresponding overall transfer unit.

The equilibrium conditions are expressed by the concentration of acid in the ketone phase being 0·548 times that in the aqueous phase.

Solution

The solvent flow rate $L'_E = 0.0014 \text{ m}^3/\text{s m}^2$

and the increase in concentration of the extract stream,

$$(C_{E_2} - C_{E_1}) = (0.38 - 0) = 0.38 \text{ kmol/m}^3$$

At the bottom of the column, $C_{R_1} = 0.82 \text{ kmol/m}^3$

and the equilibrium concentration

$$C^*_{E_1} = (0.548 \times 0.82) = 0.449 \text{ kmol/m}^3$$

$$C_{E_1} = 0$$

$$\therefore \quad \Delta C_1 = (C^*_{E_1} - C_{E_1}) = 0.449 \text{ kmol/m}^3$$

At the top of the column,

$$C_{R_2} = 1.19 \text{ kmol/m}^3$$

and hence $C^*_{E_2} = (0.548 \times 1.19) = 0.652 \text{ kmol/m}^3$

$$C_{E_2} = 0.38 \text{ kmol/m}^3$$

$$\therefore \quad \Delta C_2 = (C^*_{E_2} - C_{E_2}) = (0.652 - 0.38) = 0.272 \text{ kmol/m}^3$$

The logarithmic mean driving force

$$\Delta C_{lm} = (0\cdot449 - 0\cdot272)/\ln{(0\cdot449/0\cdot272)}$$
$$= 0\cdot353\,\text{kmol/m}^3$$

The effective height $Z = 1\cdot09\,\text{m}$

and in equation 13.37:

$$0\cdot0014\,(0\cdot38 - 0) = K_E a \times 0\cdot353 \times 1\cdot09$$

and $K_E a = \underline{\underline{0\cdot001\,38/\text{s}}}$

The height of an overall transfer unit based on concentrations in the extract phase is given by equation 13.33:

$$\mathbf{H}_{OE} = L_E'/K_E a = (0\cdot0014/0\cdot001\,38) = \underline{\underline{1\cdot02\,\text{m}}}$$

Problem 13.2

A laboratory examination is made of the extraction of acetic acid from dilute aqueous solution by means of methyl iso-butyl ketone using a spray tower of 47 mm diameter and 1080 mm high. The aqueous liquor is run into the top of the tower and the ketone enters at the bottom.

The ketone enters at a rate of $0\cdot0022\,\text{m}^3/\text{s}\,\text{m}^2$ of tower cross-section containing no acetic acid, and leaves with a concentration of $0\cdot21\,\text{kmol/m}^3$. The aqueous phase flows at a rate of $0\cdot0013\,\text{m}^3/\text{s}\,\text{m}^2$ of tower cross-section and enters containing $0\cdot68\,\text{kmol}$ acid/m³.

Calculate the overall extraction coefficient based on the driving force in the ketone phase. What is the corresponding value of the overall HTU based on the ketone phase?

Using units of kmol/m³, the equilibrium relationship under these conditions may be taken as:

Concentration of acid in the ketone phase equals 0·548 times the concentration in the aqueous phase.

Solution

The increase in concentration of the extract phase

$$(C_{E_2} - C_{E_1}) = 0\cdot21\,\text{kmol/m}^3$$

and the amount of acid transferred to the ketone phase

$$L_E'(C_{E_2} - C_{E_1}) = (0\cdot0022 \times 0\cdot21) = 0\cdot000\,462\,\text{kmol/m}^2\,\text{s}$$

Making a mass balance by way of equation 13.28:

$$0\cdot000\,462 = 0\cdot0013\,(0\cdot68 - C_{R_1})$$

$$\therefore \qquad C_{R_1} = 0\cdot325\,\text{kmol/m}^3$$

At the top of the column:

$$C_{R_2} = 0.68 \, \text{kmol/m}^3$$

$$\therefore \quad C_{E_2}^* = (0.548 \times 0.68) = 0.373 \, \text{kmol/m}^3$$

$$C_{E_2} = 0.21 \, \text{kmol/m}^3$$

$$\therefore \quad \Delta C_2 = (0.373 - 0.21) = 0.163 \, \text{kmol/m}^3$$

At the bottom of the column:

$$C_{R_1} = 0.325 \, \text{kmol/m}^3$$

$$\therefore \quad C_{E_1}^* = (0.548 \times 0.325) = 0.178 \, \text{kmol/m}^3$$

$$C_{E_1} = 0$$

$$\therefore \quad \Delta C_1 = (0.178 - 0) = 0.178 \, \text{kmol/m}^3$$

The logarithmic mean driving force is thus:

$$\Delta C_{\text{lm}} = (0.178 - 0.163)/\ln(0.178/0.163)$$

$$= 0.170 \, \text{kmol/m}^3$$

The height $Z = 1.08$ m and in equation 13.37:

$$0.0022(0.21 - 0) = K_E a \times 0.170 \times 1.08$$

$$K_E a = \underline{0.0025/\text{s}}$$

In equation 13.33: $\quad \mathbf{H}_{OE} = (0.0022/0.0025)$

$$= \underline{\underline{0.88 \, \text{m}}}$$

Problem 13.3

Propionic acid is extracted with water from a dilute solution in benzene by bubbling the benzene phase into the bottom of a tower to which water is fed at the top.

The tower is 1·2 m high and 0·14 m^2 in area, the drop volume is 0·12 cm^3, and the velocity of rise 12 cm/s. From experiments in the laboratory the value of K_W for forming drops is 6×10^{-5} kmol/s m^2 (kmol/m^3) and for rising drops $K_W = 4.2 \times 10^{-5}$ kmol/s m^2 (kmol/m^3).

What do you expect to be the value of $K_W a$ for the tower in kmol/s m^3 (kmol/m^3)?

Solution

Considering drop formation

$$K_W = 6 \times 10^{-5} \, \text{kmol/s m}^2 \, (\text{kmol/m}^3)$$

Droplet volume = 0.12 cm^3 or 1.2×10^{-7} m^3.
Radius of a drop = $(3 \times 1.2 \times 10^{-7}/4\pi)^{0.33} = 3.08 \times 10^{-3}$ m.
Mean area during formation, from Problem 13.6 = $12\pi r^2/5$

$$= 12\pi(3.08 \times 10^{-3})^2/5 = 7.14 \times 10^{-5} \, \text{m}^2$$

Mean time of exposure $= 3t_f/5$ s, where t_f, the time of formation, may be taken as $t_f = $ volume of one drop/volumetric throughput $= 1\cdot2 \times 10^{-7}/v$, where v m^3/s is the volumetric throughput of the benzene phase.

$\therefore \qquad$ mean time of exposure $= 3 \times 1\cdot2 \times 10^{-7}/5v = 7\cdot2 \times 10^{-8}/v$ s

$$\text{and mass transferred} = 6 \times 10^{-5} \times 7\cdot14 \times 10^{-5} \times 7\cdot2 \times 10^{-8}/v$$

$$= 3\cdot24 \times 10^{-16}/v \,\text{kmol}/(\text{kmol/m}^3)$$

Considering drop rise

$$K_W = 4\cdot2 \times 10^{-5} \,\text{kmol/s m}^2 \,(\text{kmol/m}^3)$$

$$\text{Residence time} = 1\cdot2/0\cdot12 = 10\,\text{s}$$

$\therefore \qquad$ volume in suspension $= 10v \,\text{m}^3$

$\therefore \qquad$ number of drops rising $= 10v/1\cdot2 \times 10^{-7}$

$$= 8\cdot3 \times 10^7 v$$

Area of one drop $= 4\pi(3\cdot08 \times 10^{-3})^2 = 1\cdot19 \times 10^{-4} \,\text{m}^2$

and interfacial area available $= 8\cdot3 \times 10^7 v \times 1\cdot19 \times 10^{-4}$

$$= 9\cdot88v \times 10^3 \,\text{m}^2$$

$\therefore \qquad$ mass transferred $= 4\cdot2 \times 10^{-5} \times 10 \times 9\cdot88 \times 10^3$

$$= 4\cdot15v \,\text{kmol}/(\text{kmol/m}^3)$$

Total mass transferred $= 4\cdot15v + 3\cdot24 \times 10^{-16}/v \,\text{kmol}/(\text{kmol/m}^3)$.
Total residence time $= (10 + 1\cdot2 \times 10^{-7}/v)$ s.
Volume of column $= (1\cdot2 \times 0\cdot14) = 0\cdot168 \,\text{m}^3$.

$\therefore \qquad K_W a = (4\cdot15v + 3\cdot24 \times 10^{-16}/v)/[0\cdot168(10 + 1\cdot2 \times 10^{-7}/v)]$

which is approximately equal to

$$\underline{2\cdot47v \,\text{kmol/s m}^3 \,(\text{kmol/m}^3)}$$

and further solution is not possible without data on the volumetric throughput of the benzene phase.

Problem 13.4

A 50% solution of solute C in solvent A is extracted with a second solvent B in a countercurrent multiple contact extraction unit. The weight of B is 25% that of the feed solution, and the equilibrium data are as given in Fig. 13a.

Determine the number of ideal stages required and the weight and concentration of the first extract if the final raffinate has 15% of solute C.

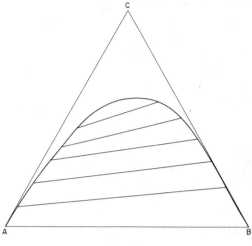

FIG. 13a

Solution

The equilibrium data are replotted in Fig. 13b and F, representing the feed, is drawn in on AC at $C = 0.50$, $A = 0.50$. FB is joined and M located such that $FM/MB = 0.25$. R_n is located on the equilibrium curve such that $C = 0.15$. (In fact $B = 0.01$ and $A = 0.84$.) E_1 is located by projecting $R_n M$ to the curve and the pole P by projecting $E_1 F$ and BR_n. R_1 is found by projecting from E_1 along a tie-line and E_2 as the projection of PR_1. The working is continued in this way and it is found that R_5 is below R_n and hence 5 ideal stages are required.

From Fig. 13b the concentration of extract E_1 is 9% A, 58% C, and 33% B.

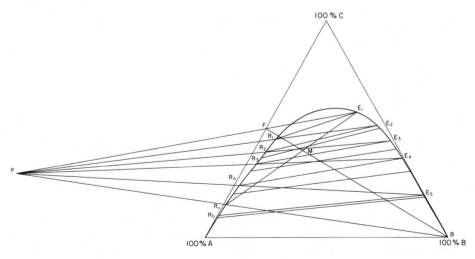

FIG. 13b

Problem 13.5

Acetaldehyde (5%) is in solution in toluene and is to be extracted with water in a five-stage co-current unit. If 25 kg of water is used per 100 kg feed, find the amount of acetaldehyde extracted and the final concentration.

If $Y =$ kg acetaldehyde/kg water and $X =$ kg acetaldehyde/kg toluene, then $Y_e = 2{\cdot}20X$ represents the equilibrium relation.

Solution

Use is made of equation 13.13 which applies to co-current contact with immiscible solvents where the equilibrium curve is a straight line as in the present case:

$$X_n = [A/(A + Bm)]^n X_f$$

On the basis of 100 kg feed:

the mass of solvent in the feed, $A = 95$ kg,
the mass of solvent added, $B = 25$ kg,
the slope of the equilibrium line, $m = 2{\cdot}20$ kg/kg,
the number of stages, $n = 5$, and
the mass ratio of solute in the feed $X_f = (5/95) = 0{\cdot}0527$ kg/kg.

$\therefore$ mass ratio of solute in raffinate

$$X_n = [95/(95 + 2{\cdot}2 \times 25)]^5 \times 0{\cdot}0527$$

$$= 0{\cdot}005\,38 \text{ kg/kg solvent}$$

Thus, the final solution consists of $0{\cdot}005\,38$ kg acetaldehyde in $1{\cdot}005\,38$ kg solution and the concentration

$$= (100 \times 0{\cdot}005\,38)/1{\cdot}005\,38$$

$$= \underline{\underline{0{\cdot}536\%}}$$

With 95 kg toluene in the raffinate,

$$\text{mass of acetaldehyde} = (0{\cdot}005\,38 \times 95)$$

$$= 0{\cdot}511 \text{ kg}$$

and $\qquad$ mass of acetaldehyde extracted $= (5{\cdot}0 - 0{\cdot}511)$

$$= \underline{\underline{4{\cdot}489 \text{ kg per 100 kg feed}}}$$

Problem 13.6

If a drop is formed in an immiscible liquid, show that the average surface available during formation of the drop is $(12/5)\,\pi r^2$, where r is the radius of the drop, and that the average time of exposure of the interface is $3t_f/5$, where t_f is the time of formation of the drop.

Solution

Assume the volumetric input to the drop is constant or,

$$4\pi r^3/3t = k$$

$$\therefore \quad r^2 = (3k/4\pi)^{2/3}t^{2/3}$$

The surface area at radius r is given by:

$$a = 4\pi r^2$$

or substituting for r^2:

$$a = 4\pi(3k/4\pi)^{2/3}t^{2/3}$$

The mean area exposed between time 0 and t is then:

$$\bar{a} = (1/t)\int_0^t a \; dt$$

$$= (1/t)\,4\pi(3k/4\pi)^{2/3}\int_0^t t^{2/3} \; dt$$

$$= 4\pi(3k/4\pi)^{2/3}(3t^{2/3}/5)$$

$$= \underline{12\pi r^2/5}$$

The area under the curve of a as a function of t is

$$\int_0^t a \; dt \quad \text{or} \quad 12\pi r^2 t/5$$

and the mean time of exposure:

$$\bar{t} = (1/a)\int_0^t a \; dt$$

$$= 12\pi r^2 t/5a$$

When $t = t_f$, the time of formation $a = 4\pi r^2$

and

$$\bar{t} = 12\pi r^2 t_f/20\pi r^2 = \underline{3t_f/5}$$

Problem 13.7

In the extraction of acetic acid from an aqueous solution with benzene in a packed column of height 1·4 m and cross-sectional area 0·0045 m², the concentrations measured at the inlet and the outlet of the column are as follows:

Acid concentration in the inlet water phase, $C_{W_2} = 0{\cdot}69$ kmol/m³.
Acid concentration in the outlet water phase, $C_{W_1} = 0{\cdot}684$ kmol/m³.
Flow rate of benzene phase $= 5{\cdot}6 \times 10^{-6}$ m³/s.
$= 1{\cdot}24 \times 10^{-3}$ m³/m² s.

Inlet benzene phase concentration, $C_{B_1} = 0.0040\,\text{kmol/m}^3$.
Outlet benzene phase concentration, $C_{B_2} = 0.0115\,\text{kmol/m}^3$.

Determine the overall transfer coefficient and the height of the transfer unit.

Solution

The acid transferred to the benzene phase is:

$$5.6 \times 10^{-6}(0.0115 - 0.0040) = 4.2 \times 10^{-8}\,\text{kmol/s}$$

The equilibrium relationship for this system is

$$C_B^* = 0.0247 C_W$$

∴ $$C_{B_1}^* = 0.0247 \times 0.684 = 0.0169\,\text{kmol/m}^3$$

and $$C_{B_2}^* = 0.0247 \times 0.690 = 0.0171\,\text{kmol/m}^3$$

∴ driving force at bottom, $\Delta C_1 = 0.0169 - 0.0040 = 0.0129\,\text{kmol/m}^3$

and driving force at top, $\Delta C_2 = 0.0171 - 0.0115 = 0.0056\,\text{kmol/m}^3$

Logarithmic mean driving force, $\Delta C_{\text{lm}} = 0.0087\,\text{kmol/m}^3$.

∴ $$K_B a = (\text{kmol transferred})/(\text{volume of packing} \times \Delta C_{\text{lm}})$$

$$= 4.2 \times 10^{-8}/(1.4 \times 0.0045 \times 0.0087)$$

$$= \underline{\underline{7.66 \times 10^{-4}\,\text{kmol/s}\,\text{m}^3\,(\text{kmol/m}^3)}}$$

and $$H_{OB} = 1.24 \times 10^{-3}/7.66 \times 10^{-4}$$

$$= \underline{\underline{1.618\,\text{m}}}$$

Problem 13.8

It is required to design a spray tower for the extraction of benzoic acid from solution in benzene.

Some experiments have been made on the rate of extraction of benzoic acid from a dilute solution in benzene to water in which the benzene phase was bubbled into the base of a 25 mm diameter column and the water fed to the top of the column. Arrangements were made to measure the rate of mass transfer during the formation of the bubbles in the water phase and during the rise of the bubbles up the column. For conditions where the drop volume was $0.12\,\text{cm}^3$ the velocity of rise $12.5\,\text{cm/s}$, the value of K_W for the period of drop formation was $0.000\,075\,\text{kmol/s}\,\text{m}^2\,(\text{kmol/m}^3)$ and for the period of rise, $0.000\,046$ in the same units.

If these conditions of drop formation and rise are reproduced in a spray tower of $1.8\,\text{m}$ height and $0.04\,\text{m}^2$ cross-sectional area, what value would you expect for the transfer coefficient $K_W a$ in $\text{kmol/s}\,\text{m}^3\,(\text{kmol/m}^3)$ where a represents the interfacial area in m^2 per unit volume of the column? The benzene phase enters at $38\,\text{cm}^3/\text{s}$.

Solution

Considering drop formation

$$K_W = 7.5 \times 10^{-5} \text{ kmol/s m}^2 (\text{kmol/m}^3)$$

Drop volume $= 0.12 \text{ cm}^3$.

$\therefore$ radius of drop $= (0.12 \times 3/4\pi)^{0.33} = 0.306 \text{ cm}$ or 0.0031 m

Surface area of drop $= 4\pi(0.0031)^2 = 1.177 \times 10^{-4} \text{ m}^2$.

As discussed in Problem 13.6, average surface area available during formation:

$$= 12\pi r^2/5 = 12(0.0031)^2/5$$

$$= 7.06 \times 10^{-5} \text{ m}^2$$

The average time of exposure of the drop (Problem 13.6) $= 3t_f/5$ where t_f is the formation time given by (volume of 1 drop)/(volumetric throughput) or

$$t_f = (0.12/38) = 0.003\,16 \text{ s}$$

$\therefore$ average time of exposure $= 3 \times 0.003\,16/5 = 1.90 \times 10^{-3} \text{ s}$

and

mass transferred during drop formation $= 7.5 \times 10^{-5} \times 7.06 \times 10^{-5} \times 1.90 \times 10^{-3}$

$$= 1.004 \times 10^{-11} \text{ kmol/(kmol/m}^3)$$

Considering rise of drops

$$K_W = 4.6 \times 10^{-5} \text{ kmol/s m}^2 (\text{kmol/m}^3)$$

Residence time is the time taken to rise 1.8 m at 12.5 cm/s or 0.125 m/s.

$\therefore$ residence time $= (1.8/0.125) = 14.4 \text{ s}$

The volumetric holdup $=$ (volumetric throughput) $\times$ (residence time)

$$= 38 \times 14.4 = 5.47 \times 10^2 \text{ cm}^3 \quad \text{or} \quad 5.47 \times 10^{-4} \text{ m}^3$$

Volume of 1 drop $= 0.12 \text{ cm}^3$ or $1.2 \times 10^{-7} \text{ m}^3$.

$\therefore$ number of drops in suspension $= 5.47 \times 10^{-4}/1.2 \times 10^{-7} = 4.56 \times 10^3$

and total surface area $= 4.56 \times 10^3 \times 1.177 \times 10^{-4} = 0.537 \text{ m}^2$

$\therefore$ mass transferred during droplet rise $= 4.6 \times 10^{-5} \times 0.537 \times 14.4$

$$= 3.55 \times 10^{-4} \text{ kmol/(kmol/m}^3)$$

Total mass transferred $= 1.004 \times 10^{-11} + 3.55 \times 10^{-4}$

$$= 3.55 \times 10^{-4} \text{ kmol/(kmol/m}^3).$$

Volume of column $= 1.8 \times 0.04 = 7.2 \times 10^{-2} \text{ m}^3$.

Total residence time $= 14.4 + 1.9 \times 10^{-3} = 14.4 \text{ s}$.

$\therefore$ $K_W a = 3.55 \times 10^{-4}/(14.4 \times 7.2 \times 10^{-2}) = \underline{\underline{3.42 \times 10^{-4} \text{ kmol/s m}^3 (\text{kmol/m}^3)}}$

Problem 13.9

It is proposed to reduce the concentration of acetaldehyde in aqueous solution from 50% to 5% by weight, by extraction with solvent S at 293 K. If a countercurrent multiple contact process is adopted and 0·025 kg/s of the solution is treated with an equal quantity of solvent, determine the number of theoretical stages required and the weight and concentration of the extract from the first stage.

The equilibrium relationship for this system at 293 K is given in Fig. 13c.

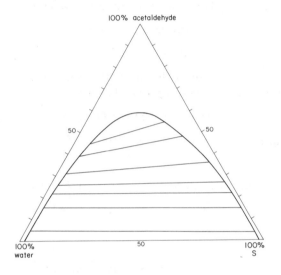

FIG. 13c

Solution

The data are replotted in Fig. 13d and the point F, representing the feed, is drawn in at 50% acetaldehyde, 50% water. Similarly, R_n, the raffinate from stage n, is located on the curve corresponding to 5% acetaldehyde. (This solution then contains 2% S and 93% water.) FS is joined and point M located such that $FM = MS$, since the ratio of feed solution to solvent is unity. $R_n M$ is projected to meet the equilibrium curve at E_1 and FE_1 and $R_n S$ are projected to meet at P. The tie-line $E_1 R_1$ is drawn in and $R_1 P$ then meets the curve at E_2. The working is continued in this way and it is found that R_4 is below R_n and hence four theoretical stages are required.

From Fig. 13d the composition of the extract from stage 1, E_1, is:

3% water, 32% acetaldehyde, and 65% S

Making an overall balance, $F + S = R_n + E_1$

∴ $0·025 + 0·025 = R_n + E_1 = 0·050 \text{ kg/s}$

Making an acetaldehyde balance:

$$0.50F + 0 = 0.05R_n + 0.32E_1$$

$$0.05(0.050 - E_1) + 0.32E_1 = (0.50 \times 0.025)$$

$$\therefore \quad E_1 = 0.037\,\text{kg/s}$$

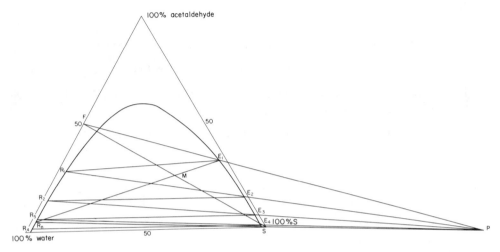

FIG. 13d

Problem 13.10

$160\,\text{cm}^3/\text{s}$ of a solvent S is used to treat $400\,\text{cm}^3/\text{s}$ of a 10% by weight solution of A in B, in a three-stage countercurrent multiple contact liquid–liquid extraction plant. What is the composition of the final raffinate?

Using the same total amount of solvent, evenly distributed between the three stages, what would be the composition of the final raffinate if the equipment were used in a simple multiple contact arrangement?

Equilibrium data:

kg A/kg B:	0.05	0.10	0.15
kg A/kg S:	0.069	0.159	0.258

Densities: $\rho_A = 1200$, $\rho_B = 1000$, $\rho_S = 800\,\text{kg/m}^3$

Solution

(1) Countercurrent operation

Considering the solvent S, $160\,\text{cm}^3/\text{s} = 1.6 \times 10^{-4}\,\text{m}^3/\text{s}$

and $\qquad$ mass flow $= (1.6 \times 10^{-4} \times 800) = 0.128\,\text{kg/s}$

Considering the solution, $400\,\text{cm}^3/\text{s} = 4 \times 10^{-4}\,\text{m}^3/\text{s}$

containing, say, $a \, \text{m}^3/\text{s} \, A$ and $(4 \times 10^{-4} - a) \, \text{m}^3/\text{s} \, B$.

$\therefore$ mass flow of $A = 1200a \, \text{kg/s}$

and   mass flow of $B = (4 \times 10^{-4} - a)1000 = (0.4 - 1000a) \, \text{kg/s}$

a total of $(0.4 + 200a) \, \text{kg/s}$

Concentration of the solution is $0.10 = 1200a/(0.4 + 200a)$

$\therefore$ $a = 3.39 \times 10^{-5} \, \text{m}^3/\text{s}$

$\therefore$ mass flow of $A = 0.041 \, \text{kg/s}$,   mass flow of $B = 0.366 \, \text{kg/s}$

$\therefore$ ratio of A/B in the feed $X_f = (0.041/0.366) = 0.112 \, \text{kg/kg}$

The equilibrium data are plotted in Fig. 13e and the value of $X_f = 0.112 \, \text{kg/kg}$ marked in. The slope of the equilibrium line is

$$\text{mass flow of } B/\text{mass flow of } S = (0.366/0.128) = 2.86$$

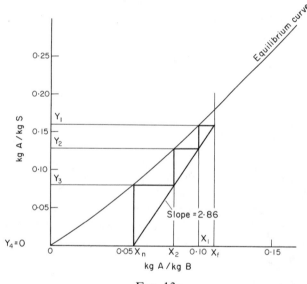

FIG. 13e

Since pure solvent is added, $Y_{n+1} = Y_4 = 0$ and a line of slope 2.86 is drawn in such that stepping off from $X_f = 0.112 \, \text{kg/kg}$ to $Y_4 = 0$ gives exactly three stages.
When $Y_4 = 0$, $X_n = X_3$ is $0.057 \, \text{kg/kg}$,

$\therefore$ composition of final raffinate is $\underline{\underline{0.057 \, \text{kg} \, A/\text{kg} \, B}}$

(2) *Multiple contact*

In this case $(0.128/3) = 0.0427 \, \text{kg/s}$ of pure solvent S is fed to each stage.

Stage 1

$$X_f = (0{\cdot}041/0{\cdot}366) = 0{\cdot}112\,\text{kg/kg}$$

and from the equilibrium curve, the extract contains $0{\cdot}18$ kg A/kg S or $(0{\cdot}18 \times 0{\cdot}0427) = 0{\cdot}0077$ kg/s A.

$\therefore$ raffinate from stage 1 contains $(0{\cdot}041 - 0{\cdot}0077) = 0{\cdot}0333$ kg/s A and $0{\cdot}366$ kg/s B

$\therefore$
$$X_1 = (0{\cdot}0333/0{\cdot}366) = 0{\cdot}091\,\text{kg/kg}$$

Stage 2

$$X_1 = 0{\cdot}091\,\text{kg/kg}$$

and from Fig. 13e the extract contains $0{\cdot}14$ kg A/kg S or $(0{\cdot}14 \times 0{\cdot}0427) = 0{\cdot}0060$ kg/s A.

$\therefore$ raffinate from stage 2 contains $(0{\cdot}0333 - 0{\cdot}0060) = 0{\cdot}0273$ kg/s A and $0{\cdot}366$ kg/s B

$\therefore$
$$X_2 = (0{\cdot}0273/0{\cdot}366) = 0{\cdot}075\,\text{kg/kg}$$

Stage 3

$$X_2 = 0{\cdot}075\,\text{kg/kg}$$

and from Fig. 13e the extract contains $0{\cdot}114$ kg A/kg S or $(0{\cdot}114 \times 0{\cdot}0427) = 0{\cdot}0049$ kg/s A.

$\therefore$ raffinate from stage 3 contains $(0{\cdot}0273 - 0{\cdot}0049) = 0{\cdot}0224$ kg/s A and $0{\cdot}366$ kg/s B

$\therefore$
$$X_3 = (0{\cdot}0224/0{\cdot}366) = 0{\cdot}061\,\text{kg/kg}$$

$\therefore$
$$\text{composition of final raffinate} = \underline{\underline{0{\cdot}061\ \text{kg}\ A/\text{kg}\ B}}$$

Problem 13.11

In order to extract acetic acid from dilute aqueous solution, with isopropyl ether, the two immiscible phases are passed countercurrently through a packed column 3 m in length and 75 mm in diameter.

It is found that if $0{\cdot}5\,\text{kg/m}^2\,\text{s}$ of the pure ether is used to extract $0{\cdot}25\,\text{kg/m}^2\,\text{s}$ of $4{\cdot}0\%$ acid by weight, then the ether phase leaves the column with a concentration of $1{\cdot}0\%$ acid by weight. Calculate:

(a) the number of overall transfer units based on the raffinate phase; and
(b) the overall extraction coefficient based on the raffinate phase.

The equilibrium relationship is given by wt. % acid in isopropyl ether phase $= 0{\cdot}3$ times the wt. % acid in the water phase.

Solution

Cross-sectional area of packing $= (\pi/4)0{\cdot}075^2 = 0{\cdot}0044\,\text{m}^2$ and volume of packing $= (0{\cdot}0044 \times 3) = 0{\cdot}0133\,\text{m}^3$.

The concentration of acid in the extract = 1.0% or 0.01 kg/kg.

∴ mass of acid transferred to the ether = $0.05(0.01 - 0) = 0.005$ kg/m² s

or $(0.005 \times 0.0044) = 0.000\,022$ kg/s

Acid in the aqueous feed = $(0.25 \times 0.04) = 0.01$ kg/m³ s.

∴ acid in raffinate = $(0.01 - 0.005) = 0.005$ kg/m² s

and concentration of acid in the raffinate = $(0.005/0.25) = 0.02$ kg/kg or 2.0%.
At the top of the column:

$$C_{R_2} = 0.040 \text{ kg/kg}$$

∴ $$C_{R_2}^* = (0.040 \times 0.3) = 0.012 \text{ kg/kg}$$

∴ $$\Delta C_2 = (0.012 - 0.010) = 0.002 \text{ kg/kg}$$

At the bottom of the column:

$$C_{R_1} = 0.20 \text{ kg/kg}$$

∴ $$C_{E_1}^* = (0.020 \times 0.3) = 0.006 \text{ kg/kg}$$

∴ $$\Delta C_1 = (0.006 - 0) = 0.006 \text{ kg/kg}$$

and the logarithmic mean driving force,

$$(\Delta C_R)_{lm} = (0.006 - 0.002)/\ln(0.006/0.002)$$

$$= 0.0036 \text{ kg/kg}$$

$$K_R a = \text{mass transferred}/[\text{volume of packing} \times (\Delta C_R)_{lm}]$$

$$= 0.000\,022/(0.0133 \times 0.0036)$$

$$= \underline{\underline{0.461 \text{ kg/m}^3 \text{ s (kg/kg)}}}$$

From equation 13.32:
the height of an overall transfer unit:

$$H_{OR} = L'_R/K_R a$$

$$= (0.25/0.461) = 0.54 \text{ m}$$

∴ number of overall transfer units = $(3/0.54) = \underline{\underline{5.53}}$

Problem 13.12

A reaction is to be carried out in an agitated vessel. Pilot plant experiments were performed under fully turbulent conditions in a tank 0·6 m in diameter, fitted with baffles and provided with a flat-bladed turbine. It was found that satisfactory mixing was obtained at a rotor speed of 4 Hz, when the power consumption was 0·15 kW and the Reynolds number 160,000. What should be the rotor speed in order to retain the same mixing performance if the linear scale of the equipment is increased 6 times? What will be the power consumption and the Reynolds number?

Solution

The correlation of power consumption and Reynolds number is given in equation 13.3 as:

$$P/D^5N^3\rho = K'(D^2N\rho/\mu)^m(DN^2/g)^n$$

In the turbulent regime with high values of Re it can be shown that **P** is independent of Re and

$$P = K'D^5N^3$$

For the *pilot-scale unit*:

$$0.15 = K'(0.6/3)^5 4^3$$

taking the impeller diameter as $D_T/3$ and hence

$$K' = 7.32 \quad \text{and} \quad P = 7.32D^5N^3 \tag{i}$$

In essence, two criteria may be used for scale-up; constant tip speed and constant power input per unit volume. Considering these in turn:

Constant impeller tip speed

The tip speed is given by $t_s = \pi DN$.

For the pilot unit, $\quad t_{s_1} = \pi \times (0.6/3) \times 4 = 2.51$ m/s

For the full-scale unit, $\quad t_{s_2} = 2.51 = \pi D_2 N_2$

or $\quad 2.51 = \pi(6 \times 0.6/3) N_2$

$\therefore \quad N_2 = 0.66$ Hz

In (i): $\quad P_2 = 7.32 D_2^5 N_2^3 = 7.35(6 \times 0.6/3)^5 0.66^3 = 5.25$ kW

For thermal similarity, that is, the same temperature in both systems, $\mu_1 = \mu_2$ and $\rho_1 = \rho_2$,

and $\quad Re_2/Re_1 = (D_2^2 N_2)/(D_1^2 N_1)$

$\therefore \quad Re_2/160,000 = [(6 \times 0.6/3)^2 \times 0.66]/[(0.6/3)^2 \times 4]$

$\therefore \quad Re_2 = 950,000$

Constant power per unit volume

Assuming the height of liquid is equal to the tank diameter, then the volume of the pilot scale unit is $(\pi/4)0.6^2 \times 0.6 = 0.170$ m^3 and the power input per unit volume = $(0.157/0.170) = 0.884$ kW/m^3.

The volume of the full-scale unit = $(\pi/4)(6 \times 0.6)^2(6 \times 0.6) = 36.6$ m^3 and hence the power consumption, $P_2 = 0.884 \times 36.6 = 32.4$ kW.

From (i): $\quad 32.4 = 7.32(6 \times 0.6/3)^5 N_2^3$

and $\quad N_2 = 1.21$ Hz

The Reynolds number is then

$$Re_2 = 160{,}000\,[(6 \times 0{\cdot}6/3)^2 \times 1{\cdot}21]/[(0{\cdot}6/3)^2 \times 4]$$

$$= 1{\cdot}74 \times 10^6$$

The choice of which scale-up criteria to adopt depends on the particular system. As a general guide constant tip speed is used where suspended solids are involved, heat is transferred to a coil or jacket, and for miscible liquids. Constant power per unit volume may be applied to immiscible liquids, emulsions, pastes, and gas–liquid systems. The former seems more appropriate in this case and hence the rotor speed should be $0{\cdot}66\,\text{Hz}$ and the power consumption will be $\underline{\underline{5{\cdot}25\,\text{kW}}}$ resulting in a Reynolds number of $\underline{\underline{9{\cdot}5 \times 10^5}}$.

Problem 13.13

It is proposed to recover material A from an aqueous effluent by washing with a solvent S and separating the resulting two phases. The light product phase will contain A and the solvent S and the heavy waste phase will contain A and water.

Show that the most economical solvent rate, W kg/s, is given by:

$$W = [(F^2 a x_0)/mb]^{0{\cdot}5} - F/m$$

where the feed rate $= F$ kg/s water containing x_0 kg A/kg water, the value of A in the solvent product phase $= £a$/kg A, the cost of solvent $S = £b$/kg S, and the equilibrium data are given by:

$$(\text{kg } A/\text{kg } S)_{\text{product phase}} = m\,(\text{kg } A/\text{kg water})_{\text{water phase}}$$

a, b, and m are constants.

Solution

The feed consists of F kg/s water containing Fx_0 kg/s A, and this is mixed with W kg/s of solvent S. The product consists of a heavy phase containing F kg/s water and, say, Fx kg/s of A, and a heavy phase containing W kg/s of S and Wy kg/s A, where y is the concentration of A in S, that is kg A/kg S.

That is,

$$y = mx$$

Making a balance in terms of the solute A,

$$Fx_0 = Wy + Fx$$

$$= Wmx + Fx = x(mW + F)$$

$$\therefore \qquad x = Fx_0/(mW + F)$$

The amount received of A recovered $= F(x_0 - x)$ kg/s

which has a value of $Fa(x_0 - x)$ £/s.

Substituting for x, the value of A recovered

$$= Fax_0 - F^2 a x_0/(mW + F) \text{ £/s}$$

The cost involved is that of the solvent used, i.e. Wb £/s.

Taking the profit P as the value of A recovered less the cost of solvents, all other costs being equal, then:

$$P = Fax_0 [1 - F/(mW + F)] - Wb \text{ £/s}$$

$$\therefore \quad dP/dW = F^2 ax_0 \, m/(mW + F)^2 - b$$

Putting the differential equal to zero for maximum profit,

$$F^2 ax_0 \, m = b(mW + F)^2$$

$$\therefore \quad \underline{\underline{W = (F^2 ax_0/mb)^{0.5} - F/m}}$$

EVAPORATION

Problem 14.1

A single-effect evaporator is used to concentrate 7 kg/s of a solution from 10 to 50% of solids. Steam is available at $205 \, kN/m^2$ and evaporation takes place at $13 \cdot 5 \, kN/m^2$.

If the overall heat transfer coefficient is $3 \, kW/m^2 \, K$, calculate the heating surface required and the amount of steam used if the feed to the evaporator is at 294 K and the condensate leaves the heating space at $352 \cdot 7 \, K$.

Specific heat of 10% solution $3 \cdot 76 \, kJ/kg \, K$.

Specific heat of 50% solution $3 \cdot 14 \, kJ/kg \, K$.

Solution

From steam tables, Vol. 1, assuming the steam is dry and saturated at $205 \, kN/m^2$, the steam temperature $= 394 \, K$ and the total enthalpy $= 2530 \, kJ/kg$.

At $13 \cdot 5 \, kN/m^2$ water boils at 325 K and in the absence of data as to the boiling-point rise, this will be taken as the temperature of evaporation, assuming an aqueous solution. The total enthalpy of steam at 325 K is $2594 \, kJ/kg$.

The feed containing 10% solids has thus to be heated from 294 to 325 K at which evaporation takes place.

In the feed, mass of dry solids $= (7 \times 10/100) = 0 \cdot 7$ kg/s and for x kg/s of water in the product:

$$0 \cdot 7 \times 100/(0 \cdot 7 + x) = 50$$

$$\therefore \qquad x = 0 \cdot 7 \, kg/s$$

$$\therefore \qquad \text{water to be evaporated} = (7 \cdot 0 - 0 \cdot 7) - 0 \cdot 7 = 5 \cdot 6 \, kg/s$$

Summarising:

	Solids (kg/s)	Liquid (kg/s)	Total (kg/s)
Feed	0·7	6·3	7·0
Product	0·7	0·7	1·4
Evaporation		5·6	5·6

Using a datum of 273 K:

Heat entering with feed $= 7 \cdot 0 \times 3 \cdot 76(294 - 273) = 552 \cdot 7 \, kW$.

Heat leaving with product $= 1 \cdot 4 \times 3 \cdot 14(325 - 273) = 228 \cdot 6 \, kW$.

Heat leaving with evaporated water $= 5\cdot6 \times 2594 = 14{,}526\cdot4\,\text{kW}$.

$\therefore$ heat transferred from steam $= (14{,}526\cdot4 + 228\cdot6) - 552\cdot7 = 14{,}202\cdot3\,\text{kW}$

The condensed steam leaves at $352\cdot7\,\text{K}$ at which the enthalpy

$$= 4\cdot18(352\cdot7 - 273) = 333\cdot2\,\text{kJ/kg}$$

Thus heat transferred from $1\,\text{kg}$ steam $= (2530 - 333\cdot2) = 2196\cdot8\,\text{kJ/kg}$ and steam required $= (14{,}202\cdot3/2196\cdot8) = \underline{\underline{6\cdot47\,\text{kg/s}}}$.

The temperature driving force will be taken as the difference between the temperature of the condensing steam and that of the evaporating water as the preheating of the solution and subcooling of the condensate represent but a small proportion of the total heat load, i.e. $\Delta T = (394 - 325) = 69\,\text{degK}$.

Thus from equation 14.1:

$$A = Q/U\Delta T$$
$$= 14{,}202\cdot3/(3 \times 69) = \underline{\underline{68\cdot6\,\text{m}^2}}$$

Problem 14.2

A solution containing 10% of caustic soda has to be concentrated to a 35% solution at the rate of 180,000 kg/day during a year of 300 working days. A suitable single-effect evaporator for this purpose (neglecting condensing plant) costs £1600 and for a multiple-effect evaporator the cost may be taken as £1600N, where N is the number of effects.

Boiler steam can be purchased at £0·2 per 1000 kg and the vapour produced may be assumed to be $0\cdot85N$ kg/kg of boiler steam. Assuming that interest on capital, depreciation, and other fixed charges amount to 45% of the capital involved per annum, and that the cost of labour is constant and independent of the number of effects employed, determine the number of effects which, on the data supplied, will give the maximum economy.

Solution

The first step is to prepare a mass balance.

Mass of caustic soda in 180,000 kg/day of feed $= 180{,}000 \times 10/100 = 18{,}000$ kg/day and mass of water in feed $= (180{,}000 - 18{,}000) = 162{,}000$ kg/day.

For x kg/day water in product, $(18{,}000 \times 100)/(18{,}000 + x) = 35$ and water in product, $x = 33{,}430$ kg/day.

$\therefore$ evaporation $= (162{,}000 - 33{,}430) = 128{,}570$ kg/day

Summarising:

	Solids (kg/day)	Liquid (kg/day)	Total (kg/day)
Feed	18,000	162,000	180,000
Product	18,000	33,430	51,430
Evaporation		128,570	128,570

The evaporation in one year is thus:

$$(128{,}570 \times 300) = 3{\cdot}86 \times 10^7 \, \text{kg}$$

The boiler steam required $= 1/0{\cdot}85N$ kg/kg of vapour produced

or $3{\cdot}86 \times 10^7/0{\cdot}85N = 4{\cdot}54 \times 10^7 N^{-1}$ kg/year

Thus the annual cost of steam $= 0{\cdot}2 \times 4{\cdot}54 \times 10^7 N^{-1}/1000$

$$= 9076 N^{-1} \, \text{£/year}$$

The capital cost of the installation $= £1600N$ and the annual capital charges $= 1600N \times 45/100 = £702N$/year.

The labour cost is independent of the number of effects and hence the total annual cost C is made up of the capital charges plus the cost of steam, or:

$$C = 720N + 9076N^{-1} \, \text{£/year}$$

$\therefore$ $dC/dN = 720 - 9076N^{-2}$

In order to minimise the costs,

$$dC/dN = 0 \quad \text{or} \quad 9076N^{-2} = 720$$

from which $N = 3{\cdot}55$

Thus N must be either 3 or 4 effects.

When $N = 3$, $C = 720 \times 3 + 9076/3 = 5185$£/year

When $N = 4$, $C = 720 \times 4 + 9076/4 = 5149$£/year

and hence 4 effects would be specified.

Problem 14.3

Saturated steam is given off from an evaporator at atmospheric pressure and is being compressed by means of saturated steam at $1135 \, \text{kN/m}^2$ in a steam jet to a pressure of $135 \, \text{kN/m}^2$. If 1 kg of the high-pressure steam compresses $1{\cdot}6$ kg of the evaporated atmospheric pressure steam, comment on the efficiency of the compressor.

Solution

The efficiency of an ejector η is given by the equation:

$$\eta = (m_1 + m_2)(H_4 - H_3)/[m_1(H_1 - H_2)]$$

where m_1 is the mass of high-pressure steam (kg), m_2 is the mass of entrained steam (kg), H_1 is the enthalpy of high-pressure steam (kJ/kg), H_2 is the enthalpy of steam after isentropic expansion in the nozzle to the pressure of the entrained vapours (kJ/kg), H_3 is the enthalpy of the mixture at the start of compression in the diffuser section (kJ/kg), and H_4 is the enthalpy of the mixture after isentropic compression to the discharge pressure (kJ/kg).

The high-pressure steam is saturated at $1135\,kN/m^2$ at which $H_1 = 2780\,kJ/kg$. If this is allowed to expand isentropically down to $101\cdot3\,kN/m^2$, from an entropy–enthalpy chart, $H_2 = 2375\,kJ/kg$ (and the dryness fraction is $0\cdot882$).

Making an enthalpy balance across the system,

$$m_1 H_1 + m_2 H_e = (m_1 + m_2) H_4$$

where H_e is the enthalpy of entrained steam. Since this is saturated at $101\cdot3\,kN/m^2$,

$$H_e = 2690\,kJ/kg \quad \text{and} \quad (1 \times 2780) + (1\cdot6 \times 2690) = (1\cdot0 + 1\cdot6)\,H_4$$

$$\therefore \qquad H_4 = 2725\,kJ/kg$$

Again assuming isentropic compression from $101\cdot3$ to $135\,kN/m^2$,

$$H_3 = 2640\,kJ/kg \quad \text{(from the chart)}$$

$$\therefore \qquad \eta = (1\cdot0 + 1\cdot6)(2725 - 2640)/[1\cdot0(2780 - 2375)]$$

$$= \underline{0\cdot55}$$

This is low, since in good design overall efficiencies approach $0\cdot75$–$0\cdot80$. Obviously the higher the efficiency the greater the entrainment ratio or the higher the saving in live steam. The low efficiency is borne out by examination of Fig. 14.13b, Vol. 2, which applies to $1135\,kN/m^2$ operating pressure.

Since the pressure of entrained vapour $= 101\cdot3\,kN/m^2$ and the discharge pressure $= 135\,kN/m^2$, the required flow of live steam $= 0\cdot5\,kg/kg$ entrained vapour.

In this case the ratio is $(1\cdot0/1\cdot6) = \underline{\underline{0\cdot63\,kg/kg}}$.

Problem 14.4

A single-effect evaporator operates at $13\,kN/m^2$. What will be the heating surface necessary to concentrate $1\cdot25\,kg/s$ of 10% caustic soda to 41%, assuming a value of U of $1\cdot25\,kW/m^2\,K$, using steam at $390\,K$? The heating surface is $1\cdot2\,m$ below the liquid level.

Boiling-point rise of solution $= 30\,degK$.
Feed temperature $= 291\,K$.
Specific heat of the feed $= 4\cdot0\,kJ/kg\,K$.
Specific heat of the product $= 3\cdot26\,kJ/kg\,K$.
Specific gravity of boiling liquid $= 1\cdot39$.

Solution

Making a mass balance, in $1\cdot25\,kg/s$ feed, mass of caustic soda

$$= (1\cdot25 \times 10/100) = 0\cdot125\,kg/s$$

For $x\,kg/s$ water in the product,

$$(100 \times 0\cdot125)/(x + 0\cdot125) = 41\cdot0$$

and water in product, $x = 0\cdot180\,kg/s$.

Thus:

	Solids (kg/s)	Liquid (kg/s)	Total (kg/s)
Feed	0·125	1·125	1·250
Product	0·125	0·180	0·305
Evaporation		0·945	0·945

At a pressure of 13 kN/m², from steam tables (Vol. 1), water boils at 324 K. Thus at the surface of the liquid the temperature will be (324 + 30) = 354 K. The pressure due to the hydrostatic head of liquid = (1·2 × 1390 × 9·81)/1000 = 16·4 kN/m² and hence the pressure at the heating surface = (16·4 + 13) = 29·4 kN/m² at which the temperature of saturated steam = 341 K.

Thus the temperature at which the liquid boils at the heating surface is 371 K at which the enthalpy of steam = 2672 kJ/kg.

∴   temperature difference $\Delta T = (390 - 371) = 19 \deg K$

The heat load Q is the heat in the vapour plus the enthalpy of the product minus the enthalpy of the feed, or

$$Q = (0.945 \times 2672) + [0.305 \times 3.26(371 - 273)] - [1.250 \times 4.0(291 - 273)]$$

assuming the product is withdrawn at 371 K.

∴ $Q = 2525 + 97.4 - 90 = 2532.4 \, kW$

Thus in equation 14.1: $A = 2532.4/(1.25 \times 19)$

$$= \underline{\underline{106.6 \, m^2}}$$

Problem 14.5

Distilled water is produced from sea water by evaporation in a single-effect evaporator working on the vapour compression system. The vapour produced is compressed by a mechanical compressor of 50% efficiency, and then returned to the calandria of the evaporator. Extra steam, dry, and saturated at 650 kN/m², is bled into the steam space through a throttling valve. The distilled water is withdrawn as condensate from the steam space. Half the sea water is evaporated in the plant. The energy supplied in addition to that necessary to compress the vapour may be assumed to appear as super-heat in the vapour.

Using the data given below, calculate the quantity of extra steam required in kg/s.

Production of distillate, 0·125 kg/s.

Pressure in vapour space, atmospheric.

Temperature difference from steam to liquor, 8 degK.

Boiling-point rise of sea water, 1·1 degK.

Specific heat of sea water, 4·18 kJ/kg K.

Sea water enters the evaporator at 344 K from an external heater.

Solution

The pressure in the vapour space is $101.3 \, kN/m^2$ at which pressure water boils at 373 K. The sea water is therefore boiling at 374.1 K and the temperature in the steam space is $(374.1 + 8) = 382.1$ K. At this temperature, steam is saturated at $120 \, kN/m^2$ and has sensible and total enthalpies of 439 and 2683 kJ/kg respectively.

Making a *mass balance*, there are two inlet streams: the extra steam, say x kg/s, and the sea water feed, say y kg/s. The two outlet streams are the distilled water product, 0.125 kg/s, and the concentrated sea water, 0.5y kg/s.

$$\therefore \qquad\qquad x + y = 0.125 + 0.5y$$

or
$$x + 0.5y = 0.125 \qquad\qquad\qquad \text{(i)}$$

Making an *energy balance*, energy enters as compression power, in the steam and inlet sea water, and leaves via the sea water plus the product.

At $650 \, kN/m^2$, total enthalpy of steam $= 2761$ kJ/kg. Thus energy in this stream

$$= 2761x \, kW$$

The sea water enters at 344 K.

$$\therefore \qquad\qquad \text{enthalpy of feed} = y \times 4.18(344 - 273) = 296.8y \, kW$$

The sea water leaves the plant at 374.1 K and hence the enthalpy of concentrated sea water $= 0.5y \times 4.18(374.1 - 273) = 211.3y \, kW$.

The product has an enthalpy of 439 kJ/kg or $(439 \times 0.125) = 54.9 \, kW$.

Making a balance,
$$W + 2761x + 296.8y = 211.3y + 54.9$$

$$\therefore \qquad\qquad W + 2761x + 85.5y = 54.9 \qquad\qquad\qquad \text{(ii)}$$

where W kW is the energy supplied to the compressor.

Substituting from (i) into (ii):

$$W + 2761x + 85.5(0.25 - 2x) = 54.9$$

$$\therefore \qquad\qquad W + 2590x = 33.5 \qquad\qquad\qquad \text{(iii)}$$

For isentropic compression, the work done is given by equation 6.36, Vol. 1.

$$-W'm = [P_1 V_1/(\gamma - 1)][(P_2/P_1)^{\gamma - 1/\gamma} - 1]$$

In the compressor, 0.5y kg/s vapour is compressed from $101.3 \, kN/m^2$, the pressure in the vapour space, to $120 \, kN/m^2$, the pressure in the calandria.

At $101.3 \, kN/m^2$ and 374.1 K, the density of steam is

$$(18/22.4)(273/374.1) = 0.586 \, kg/m^3$$

and hence
$$V_1 = 0.5y/0.586 = 0.853y \, m^3/s$$

Taking $\gamma = 1.3$ for steam:

$$W' \times 0.5y = [(101.3 \times 0.853y)/(1.3 - 1)][(120/101.3)^{0.3/1.3} - 1]$$

$$0.5W'y = 288.0y(1.185^{0.231} - 1)$$

$$= 11.5y$$

$\therefore$ $W' = 23 \cdot 0 \, \text{kW}/(\text{kg/s})$

As the compressor is 50% efficient:

$$W = W'/0 \cdot 50 = 46 \cdot 0 \, \text{kW}/(\text{kg/s})$$

$$= 46 \cdot 0 \times 0 \cdot 5y = 23 \cdot 0y \, \text{kW}$$

Substituting in (ii):

$$W + 2761x + 85 \cdot 5(W/23 \cdot 0) = 54 \cdot 9$$

$\therefore$ $2761x = 54 \cdot 9 + 4 \cdot 72W$ (iv)

From (iii), $W = 33 \cdot 5 - 2590x$

and in (iv) $2761x = 54 \cdot 9 + 4 \cdot 72(33 \cdot 5 - 2590x)$

From which $\underline{x = 0 \cdot 014 \, \text{kg/s}}$

Problem 14.6

It is claimed that a jet booster requires $0 \cdot 06 \, \text{kg/s}$ of dry and saturated steam at $700 \, \text{kN/m}^2$ to compress $0 \cdot 125 \, \text{kg/s}$ of dry and saturated vapour from $3 \cdot 5 \, \text{kN/m}^2$ to $14 \cdot 0 \, \text{kN/m}^2$. Show whether this claim is reasonable.

Solution

Using the nomenclature of Problem 14.3,

$$H_1 = 2765 \, \text{kJ/kg}$$

and assuming isentropic expansion to $3 \cdot 5 \, \text{kN/m}^2$, from the entropy–enthalpy chart:

$$H_2 = 2015 \, \text{kJ/kg}$$

Making an enthalpy balance across the system, noting that the enthalpy of saturated steam at $3 \cdot 5 \, \text{kN/m}^2$ is $2540 \, \text{kJ/kg}$,

$$(0 \cdot 06 \times 2765) + (0 \cdot 125 \times 2540) = (0 \cdot 06 + 0 \cdot 125) H_4$$

$\therefore$ $H_4 = 2612 \, \text{kJ/kg}$

Assuming isentropic compression to $14 \cdot 0 \, \text{kN/m}^2$ and $H_4 = 2612 \, \text{kJ/kg}$ from $3 \cdot 5 \, \text{kN/m}^2$, then $H_3 = 2420 \, \text{kJ/kg}$ (again using the chart).
Using the equation for efficiency given in Problem 14.3,

$$\eta = (0 \cdot 06 + 0 \cdot 125)(2612 - 2420)/[0 \cdot 06(2765 - 2015)]$$

$$= \underline{0 \cdot 79}$$

As stated in the solution to Problem 14.3, with good design, overall efficiencies approach $0 \cdot 75 - 0 \cdot 80$ which just covers the value obtained. The claim is therefore reasonable.

[In Problems 14.3 and 14.6 use has been made of entropy–enthalpy diagrams for convenience. The change in enthalpy due to isentropic compression or expansion may be calculated, however, using equations 6.35 and 6.36 in Vol. 1.]

Problem 14.7

A forward-feed double-effect evaporator, having $10\,\text{m}^2$ of heating surface in each effect, is used to concentrate $0.4\,\text{kg/s}$ of caustic soda solution from 10% to 50% by weight. During a particular run, when the feed is at 328 K, the pressures in the two calandrias are 375 and $180\,\text{kN/m}^2$ respectively, while the condenser operates at $15\,\text{kN/m}^2$. Under these conditions, find:

(a) the load on the condenser;
(b) the steam economy; and
(c) the overall heat transfer coefficient in each effect.

Would there have been any advantages in using backward feed in this case?

Data

Assume negligible heat losses to the surroundings.
Physical properties of caustic soda solutions:

Per cent solids (w/w %)	Boiling-point rise (degK)	Specific heat (kJ/kg K)	Heat of dilution (kJ/kg)
10	1·6	3·85	0
20	6·1	3·72	2·3
30	15·0	3·64	9·3
50	41·6	3·22	220

Solution

A mass balance may be made as follows:

	Solids (kg/s)	Liquor (kg/s)	Total (kg/s)
Feed	0·04	0·36	0·40
Product	0·04	0·04	0·08
Evaporation		0·32	0·32

$$\therefore \qquad (D_1 + D_2) = 0.32\,\text{kg/s}$$

From the given data:

at $375\,\text{kN/m}^2$, $\quad T_0 = 414.5\,\text{K}, \quad \lambda_0 = 2141\,\text{kJ/kg}$
at $180\,\text{kN/m}^2$, $\quad T_1 = 390\,\text{K}, \quad \lambda_1 = 2211\,\text{kJ/kg}$
at $15\,\text{kN/m}^2$, $\quad T_2 = 328\,\text{K}, \quad \lambda_2 = 2370\,\text{kJ/kg}$

Making a heat balance around stage 1,

$$D_0 \lambda_0 = WC_{p_1}(T_1' - T_f) + Wh_1 + D_1 \lambda_1 \qquad \text{(i)}$$

where C_{p_1} is the mean specific heat between T_1' and T_f, $T_1' = T_1 +$ the boiling-point rise in stage 1, and h_1 is the difference in the heat of dilution between the concentration in the first effect and the feed concentration.

Similarly, around stage 2,

$$D_1 \lambda_1 + (W - D_1) C_{p_2} (T_1' - T_2') = D_2 \lambda_2 + (W - D_1) h_2 \qquad \text{(ii)}$$

It is now necessary to select values of D_1 and D_2 such that a balance is obtained in equation (ii). Try $D_1 = 0{\cdot}17$ kg/s. Therefore $D_2 = 0{\cdot}15$ kg/s.

$\therefore$ concentration of solids in the first effect $= 0{\cdot}04/[0{\cdot}04 + (0{\cdot}36 - 0{\cdot}17)]$

$$= 0{\cdot}174 \, \text{kg/kg}$$

at which, the boiling-point rise $= 5{\cdot}0$ degK
the specific heat $= 3{\cdot}75$ kJ/kg K
and the heat of dilution $= 1{\cdot}6$ kJ/kg

$\therefore$
$$T_1' = (390 + 5{\cdot}0) = 395 \, \text{K}$$

$$T_2' = (328 + 41{\cdot}6) = 369{\cdot}6 \, \text{K}$$

$$C_{p_2} = (3{\cdot}75 + 3{\cdot}22)/2 = 3{\cdot}49 \, \text{kJ/kg K}$$

The heat of dilution to be provided in the second effect $= (220 - 1{\cdot}6) = 218{\cdot}4$ kJ/kg. Thus in equation (ii),

$$2211 \times 0{\cdot}17 + (0{\cdot}4 - 0{\cdot}17) 3{\cdot}49 (395 - 369{\cdot}6) = 0{\cdot}15 \times 2370 + (0{\cdot}4 - 0{\cdot}17) 218{\cdot}4$$

or
$$396{\cdot}3 = 405{\cdot}7$$

which is close enough for the purposes of this calculation, and hence the load on the condenser, $\underline{D_2 = 0{\cdot}15 \, \text{kg/s}}$.

For the first effect,

$$C_{p_1} = (3{\cdot}85 + 3{\cdot}75)/2 = 3{\cdot}80 \, \text{kJ/kg K}$$

$$h_1 = (1{\cdot}6 - 0) = 1{\cdot}6 \, \text{kJ/kg}$$

and in (i),

$$2141 D_0 = 0{\cdot}4 \times 3{\cdot}80 (395 - 328) + 0{\cdot}4 \times 1{\cdot}6 + 0{\cdot}17 \times 2211$$

$\therefore$
$$D_0 = 0{\cdot}23 \, \text{kg/s}$$

and the economy is $(0{\cdot}36/0{\cdot}23) = \underline{\underline{1{\cdot}57}}$

The overall coefficient in the first effect is given by:

$$U_1 = D_0 \lambda_0 / A_1 \Delta T_1$$

$$= (0{\cdot}23 \times 2141)/[10(414{\cdot}5 - 395)] = \underline{\underline{2{\cdot}53 \, \text{kW/m}^2 \, \text{K}}}$$

and for the second effect,

$$U_2 = (0{\cdot}17 \times 2211)/10(390 - 369{\cdot}6) = \underline{\underline{1{\cdot}84 \, \text{kW/m}^2 \, \text{K}}}$$

With a *backward feed* arrangement, the concentration of solids in the first effect would be 50%, which gives a boiling-point rise of 41·6 degK. The vapour passing to the second effect must condense at 390 K in the second effect calandria to give a pressure there of $180 \, kN/m^2$. Thus the temperature of the liquor in the first effect would be $(390 + 41 \cdot 6) = 431 \cdot 6$ K, which is higher than the feed steam temperature, 414·5, and thus there is no temperature gradient.

In this way there is no advantage in using backward feed.

Problem 14.8

A 12% glycerol–water mixture is produced as a secondary product in a continuous process plant and flows from the reactor at $4 \cdot 5 \, MN/m^2$ and at a temperature of 525 K. Suggest, with preliminary calculations, a method of concentration to 75% glycerol in a factory which has no low-pressure steam available.

Solution

Making a mass balance on the basis of 1 kg feed:

	Glycerol (kg)	Water (kg)	Total (kg)
Feed	0·12	0·88	1·00
Product	0·12	0·04	0·16
Evaporation	—	0·84	0·84

Thus the total evaporation is 0·84 kg water/kg feed.

The first possibility is to take account of the fact that the feed is at high pressure in that it could be reduced to, say, atmospheric pressure and the water removed by flash evaporation. For this to be possible, the heat content of the feed must be at least equal to the latent heat of the water evaporated.

Assuming evaporation at $101 \cdot 3 \, kN/m^2$, that is at 373 K and a specific heat for a 12% glycerol solution of $4 \cdot 0 \, kJ/kg \, K$, then

$$\text{heat in feed} = 1 \cdot 0 \times 4 \cdot 0 (525 - 375) = 608 \, kJ$$

$$\text{heat in water evaporated} = 0 \cdot 84 \times 2256 = 1895 \, kJ$$

and hence only $608/2256 = 0 \cdot 27$ kg water could be evaporated by this means, giving a solution containing $100 \times 0 \cdot 12/[0 \cdot 12 + (0 \cdot 88 - 0 \cdot 27)]$

$$= 16 \cdot 5\% \text{ glycerol}$$

It is therefore necessary to provide an additional source of heat. Although low-pressure steam is not available, presumably a high-pressure supply (say $1135 \, kN/m^2$) exists, and thought may be given to vapour recompression using a steam-jet ejector.

Assume that the discharge pressure from the ejector, that is in the steam chest, is

170 kN/m^2 at which the latent heat $\lambda_0 = 2216$ kN/m^2, then a heat balance across the unit gives:

$$D_0 \lambda_0 = WC_p(T_1 - T_f) + D_1 \lambda_1$$

$$\therefore \qquad 2216D_0 = 1{\cdot}0 \times 4{\cdot}0(373 - 525) + 0{\cdot}84 \times 2256$$

$$\therefore \qquad D_0 = 0{\cdot}58 \, \text{kg}$$

Using Fig. 14.13b (Vol. 2), with a pressure of entrained vapour of 101·3 kN/m^2, a live-steam pressure of 1135 kN/m^2 and a discharge pressure of 170 kN/m^2, then 1·5 kg live steam is required/kg of entrained vapour.

Thus if x kg is the amount of entrained vapour,

$$(1 + 1{\cdot}5x) = 0{\cdot}58$$

$$\therefore \qquad x = 0{\cdot}23 \, \text{kg}$$

In this way the proposal is to condense $(0{\cdot}84 - 0{\cdot}23) = 0{\cdot}61$ kg of the water evaporated and to entrain 0·23 kg with $(1{\cdot}5 \times 0{\cdot}23) = 0{\cdot}35$ kg of steam at 1135 kN/m^2 in an ejector to provide 0·58 kg of steam at 170 kN/m^2 which is fed to the calandria.

These values represent only one solution to the problem and variation of the calandria and live-steam pressures may result in even lower requirements of high-pressure steam.

Problem 14.9

A forward-feed double-effect standard vertical evaporator with equal heating areas in each effect is fed with 5 kg/s of a liquor of specific heat 4·18 kJ/kg K, and with no boiling-point rise, so that half of the feed liquor is evaporated. The overall heat transfer coefficient in the second effect is three-quarters of that in the first. Steam is fed in at 395 K and the boiling-point in the second effect is 373 K. The feed is heated by an external heater to the boiling-point in the first effect.

It is decided to bleed off 0·25 kg/s of vapour from the vapour line to the second effect for use in another process. If the feed is still heated to the boiling-point of the first effect by external means, what will be the change in the steam consumption of the evaporator unit?

For the purposes of calculation, the latent heat of the vapours and of the live steam may be taken as 2230 kJ/kg.

Solution

The total evaporation $(D_1 + D_2) = 5{\cdot}0/2 = 2{\cdot}5$ kg/s.

In equation 14.7,

$$U_1 A_1 \Delta T_1 = U_2 A_2 \Delta T_2$$

Since $\qquad A_1 = A_2 \quad$ and $\quad U_2 = 0{\cdot}75U_1$

$$\therefore \qquad \Delta T_1 = 0{\cdot}75\Delta T_2$$

Also $\qquad \Sigma \Delta T = \Delta T_1 + \Delta T_2 = (395 - 373) = 22 \, \text{degK}$

and solving between these two equations,

$$\Delta T_1 = 9 \cdot 5 \, \text{degK} \quad \text{and} \quad \Delta T_2 = 12 \cdot 5 \, \text{degK}$$

$$\therefore \qquad \text{for steam to the first effect}, \ T_0 = 395 \, \text{K}$$

$$\text{for steam to the second effect}, \ T_1 = 385 \cdot 5 \, \text{K} \quad \text{and} \quad T_2 = 373 \, \text{K}$$

λ being 2230 kJ/kg in each case.

Making a heat balance across the first effect,

$$D_0 \lambda_0 = W C_p (T_1 - T_f) + D_1 \lambda_1$$

or

$$2230 D_0 = 5 \cdot 0 \times 4 \cdot 18 (385 \cdot 5 - 385 \cdot 5) + D_1 \, 2230$$

$$\therefore \qquad D_0 = D_1$$

Making a heat balance across the second effect,

$$D_1 \lambda_1 + (W - D_1) C_p (T_1 - T_2) = D_2 \lambda_2$$

$$\therefore \qquad 2230 D_1 + (5 \cdot 0 - D_1) 4 \cdot 18 (385 \cdot 5 - 273) = 2230 D_2$$

$$\therefore \qquad 2178 D_1 = 2230 D_2 - 261 \cdot 3$$

But

$$D_2 = 2 \cdot 5 - D_1$$

$$\therefore \qquad D_1 = 1 \cdot 21 \, \text{kg/s}$$

and the steam consumption, $\qquad \underline{D_0 = 1 \cdot 21 \, \text{kg/s}}$

Considering the case where 0·25 kg/s is bled from the steam line to the calandria of the second effect, a heat balance across the first effect gives

$$D_0 = D_1 \quad \text{as before.}$$

Making a heat balance across the second effect,

$$(D_1 - 0 \cdot 25) \lambda_1 + (W - D_1) C_p (T_1 - T_2) = D_2 \lambda_2$$

$$\therefore \qquad 2230 (D_1 - 0 \cdot 25) + (5 \cdot 0 - D_1) 4 \cdot 18 (395 \cdot 5 - 373) = 2230 D_2$$

$$\therefore \qquad 2150 \cdot 7 D_1 = 2230 D_2 + 296 \cdot 2$$

Substituting $(2 \cdot 5 - D_1)$ for D_2,

$$D_1 = 1 \cdot 33 \, \text{kg/s}$$

and the steam consumption, $\qquad \underline{D_0 = 1 \cdot 33 \, \text{kg/s}}$

The change in steam consumption is therefore an increase of $\underline{\underline{0 \cdot 12 \, \text{kg/s}.}}$

Problem 14.10

A liquor containing 15% solids is concentrated to 55% solids in a double-effect evaporator operating at a pressure in the second effect of 18 kN/m². No crystals are formed. The feed is 2·5 kg/s at a temperature of 375 K with a specific heat of 3·75 kJ/kg K.

The boiling-point rise of the concentrated liquor is 6 degK and the steam fed to the first effect is at 240 kN/m².

The overall heat transfer coefficients in the first and second effects are 1·8 and 0·63 kW/m² K respectively. If the heat transfer area is to be the same in each effect, determine its value.

Solution

Making a mass balance based on a flow of feed of 2·5 kg/s,

	Solids (kg/s)	Liquor (kg/s)	Total (kg/s)
Feed	0·375	2·125	2·50
Product	0·375	0·307	0·682
Evaporation	—	1·818	1·818

$$\therefore \qquad\qquad D_1 + D_2 = 1\text{·}818 \text{ kg/s} \qquad\qquad \text{(i)}$$

At 18 kN/m² pressure, $T_2 = 331$ K and T_2', the temperature of the liquor in the second effect = 337 K allowing for the boiling-point rise.

At 240 kN/m² pressure, $T_0 = 399$ K.

$$\therefore \qquad\qquad \Delta T_1 + \Delta T_2 = (399 - 337) = 62 \text{ degK} \qquad\qquad \text{(ii)}$$

From equation 14.8, $U_1 \Delta T_1 = U_2 \Delta T_2$

Substituting $U_1 = 1\text{·}8$ and $U_2 = 0\text{·}63$ kW/m² K and combining with (ii),

$$\Delta T_1 = 16 \text{ degK} \quad \text{and} \quad \Delta T_2 = 46 \text{ degK}$$

Thus $T_1 = (399 - 16) = 383$ K and hence the feed enters slightly cooler than the temperature of the liquor in the first effect. Hence ΔT_1 will be slightly greater and the following values will be assumed:

$$\Delta T_1 = 17 \text{ degK}, \quad \Delta T_2 = 45 \text{ degK}$$

Thus, for steam to 1: $T_0 = 399$ K, $\lambda_0 = 2185$ kJ/kg
      for steam to 2: $T_1 = 382$ K, $\lambda_1 = 2232$ kJ/kg
and $T_2 = 331$ K, $\lambda_2 = 2363$ kJ/kg

Making a heat balance around each effect:

(1) $D_0 \lambda_0 = W C_p (T_1 - T_f) + D_1 \lambda_1$ or $2185D_0 = 2\text{·}5 \times 3\text{·}75(382 - 375) + 2232D_1$ (iii)

(2) $D_1 \lambda_1 + (W - D_1) C_p (T_1 - T_2') = D_2 \lambda_2$

$$\text{or} \quad 2232D_1 + (2\text{·}5 - D_1) 3\text{·}75(382 - 337) = 2363D_2 \qquad\qquad \text{(iv)}$$

Solving equations (i), (iii), and (iv) simultaneously,

$$D_0 = 0\text{·}924 \text{ kg/s}, \quad D_1 = 0\text{·}875 \text{ kg/s}, \quad \text{and} \quad D_2 = 0\text{·}943 \text{ kg/s}$$

The areas are given by:

$$A_1 = D_0 \lambda_0 / U_1 \Delta T_1 = (0.924 \times 2185)/(1.8 \times 17) = 66.1 \, \text{m}^2$$

$$A_2 = D_1 \lambda_1 / U_2 \Delta T_2 = (0.875 \times 2232)/(0.63 \times 45) = 68.8 \, \text{m}^2$$

which are sufficiently close to justify the assumed values of ΔT_1 and ΔT_2. A total area of $134.9 \, \text{m}^2$ is indicated and hence an area of $67.5 \, \text{m}^2$ would be specified for each effect.

Problem 14.11

Liquor containing 5% solids is fed at 340 K to a four-effect evaporator. Forward feed is used to give a product containing 28.5% solids. Do the following figures indicate normal operation? If not, suggest a reason for the abnormality.

Effect	I	II	III	IV
Per cent solids entering	5.0	6.6	9.1	13.1
Temperature in steam chest (K)	382	374	367	357.5
Temperature of boiling solution (K)	369.5	364.5	359.6	336.6

Solution

Examination of the data indicates one immediate point in that the temperatures in the steam chests in effects II and III are higher than the temperatures of the boiling solution in the previous effects. The explanation for this is not clear although a steam leak in the previous effect is a possibility. Further calculations may be made as follows, starting with a mass balance on the basis of 1 kg feed.

	Solid (kg)	Liquor (kg)	Total (kg)	
Feed	0.05	0.950	1.00	
Product from I	0.05	0.708	0.758	$\therefore D_1 = 0.242 \, \text{kg}$
Product from II	0.05	0.500	0.550	$\therefore D_2 = 0.208 \, \text{kg}$
Product from III	0.05	0.332	0.382	$\therefore D_3 = 0.168 \, \text{kg}$
Product from IV	0.05	0.125	0.175	$\therefore D_4 = 0.207 \, \text{kg}$

and the total evaporation = $0.825 \, \text{kg}$.

The steam fed to the plant may be obtained by a heat balance across I:

$$D_0 \lambda_0 = W C_p (T_1 - T_f) + D_1 \lambda_1$$

Taking C_p as $4.18 \, \text{kJ/kg K}$ and $\lambda_0 = \lambda_1$ as $2300 \, \text{kJ/kg}$ in all effects,

$$2300 D_0 = 1.0 \times 4.18(369.5 - 340) + 2300 \times 0.242$$

$$\therefore \qquad D_0 = 0.296 \, \text{kg}$$

The overall coefficient in each effect assuming equal areas, A m^2, is:

$$U_1 = D_0 \lambda_0/A\Delta T_1 = (0\cdot296 \times 2300)/12\cdot5A = 54\cdot5/A \, \text{kW/m}^2 \, \text{K}$$

$$U_2 = D_1 \lambda_1/A\Delta T_2 = (0\cdot242 \times 2300)/9\cdot5A = 58\cdot6/A \, \text{kW/m}^2 \, \text{K}$$

$$U_3 = D_2 \lambda_2/A\Delta T_3 = (0\cdot208 \times 2300)/6\cdot6A = 72\cdot5/A \, \text{kW/m}^2 \, \text{K}$$

$$U_4 = D_3 \lambda_3/A\Delta T_4 = (0\cdot168 \times 2300)/20\cdot9A = 18\cdot5/A \, \text{kW/m}^2 \, \text{K}$$

These results are surprising in that a reduction in U is normally obtained with decrease in boiling temperature. On this basis U_3 is high, which may indicate a change in boiling mechanism although ΔT_3 is reasonable. Even more important is the very low value of U in effect IV. This must surely indicate that part of the area is inoperative, possibly due to the formation of crystals at the high concentration.

Problem 14.12

1·25 kg/s of a solution is concentrated from 10% to 50% solids in a triple-effect evaporator using steam at 393 K and a vacuum such that the boiling-point in the last effect is 325 K. If the feed is initially at 297 K and a backward feed is used, what will be the steam consumption, the temperature distribution in the system, and the heat transfer area in each effect, each effect being identical?

For the purposes of calculation, assume that the specific heat is 4·18 kJ/kg K over the temperature range in the system. The overall heat transfer coefficient may be taken as 2·5, 2·0, and 1·6 kW/m^2 K in the first, second, and third effects.

Solution

Making a mass balance:

	Solid (kg/s)	Liquor (kg/s)	Total (kg/s)
Feed	0·125	1·125	1·250
Product	0·125	0·125	0·250
Evaporation	—	1·0	1·0

$\therefore$

$$D_1 + D_2 + D_3 = 1\cdot0 \, \text{kg/s} \tag{i}$$

$$\Sigma\Delta T = (T_0 - T_3) = (393 - 325) = 68 \, \text{degK} \tag{ii}$$

From equation 14·8, $2\cdot5\Delta T_1 = 2\cdot0\Delta T_2 = 1\cdot6\Delta T_3$ \hfill (iii)

and from (ii) and (iii),

$$\Delta T_1 = 18 \, \text{degK}, \quad \Delta T_2 = 22 \, \text{degK}, \quad \text{and} \quad \Delta T_3 = 28 \, \text{degK}$$

Modifying the figures to take account of the feed temperature, it will be assumed that:

$$\Delta T_1 = 19 \, \text{degK}, \quad \Delta T_2 = 24 \, \text{degK}, \quad \text{and} \quad \Delta T_3 = 25 \, \text{degK}$$

The temperatures in each effect and the corresponding latent heats are:

$$T_0 = 393\,K, \quad \lambda_0 = 2202\,kJ/kg$$
$$T_1 = 374\,K, \quad \lambda_1 = 2254\,kJ/kg$$
$$T_2 = 350\,K, \quad \lambda_2 = 2315\,kJ/kg$$
$$T_3 = 325\,K, \quad \lambda_3 = 2376\,kJ/kg$$

Writing a heat balance for each effect:

(1) $D_0\lambda_0 = (W - D_3 - D_2)C_p(T_1 - T_2) + D_1\lambda_1$

$$\text{or} \quad 2202D_0 = (1{\cdot}25 - D_2 - D_3)4{\cdot}18(374 - 350) + 2254D_1 \quad \text{(iv)}$$

(2) $D_1\lambda_1 = (W - D_3)C_p(T_2 - T_3) + D_2\lambda_2$

$$\text{or} \quad 2254D_1 = (1{\cdot}25 - D_3)4{\cdot}18(350 - 325) + 2315D_2 \quad \text{(v)}$$

(3) $D_2\lambda_2 = WC_p(T_3 - T_f) + D_3\lambda_3$

$$\text{or} \quad 2315D_2 = 1{\cdot}25 \times 4{\cdot}18(325 - 297) + 2376D_3 \quad \text{(vi)}$$

Solving equations (i), (iv), (v), and (vi) simultaneously,

$$D_0 = 0{\cdot}432\,kg/s, \quad D_1 = 0{\cdot}393\,kg/s, \quad D_2 = 0{\cdot}339\,kg/s, \quad \text{and} \quad D_3 = 0{\cdot}268\,kg/s$$

The areas of transfer surface are:

$$A_1 = D_0\lambda_0/U_1\,\Delta T_1 = (0{\cdot}432 \times 2202)/(2{\cdot}5 \times 19) = 20{\cdot}0\,m^2$$
$$A_2 = D_1\lambda_1/U_2\,\Delta T_2 = (0{\cdot}393 \times 2254)/(2{\cdot}0 \times 24) = 18{\cdot}5\,m^2$$
$$A_3 = D_2\lambda_2/U_3\,\Delta T_3 = (0{\cdot}268 \times 2315)/(1{\cdot}6 \times 25) = 15{\cdot}5\,m^2$$

which are probably sufficiently close for design purposes.
The steam consumption is therefore,

$$D_0 = \underline{\underline{0{\cdot}432\,kg/s}}$$

The temperatures in each effect are:

$$\underline{(1)\ 374\,K, \quad (2)\ 350\,K, \quad \text{and} \quad (3)\ 325\,K}$$

The area in each effect, taken as the mean of the calculated values = $\underline{\underline{18{\cdot}0\,m^2}}$.

Problem 14.13

A triple-effect evaporator concentrates a liquid with no appreciable elevation of boiling-point. If the temperature of the steam to the first effect is 395 K and vacuum is applied to the third effect so that the boiling-point is 325 K, what are the approximate boiling-points in the three effects? The overall transfer coefficients may be taken as 3·1, 2·3, and 1·1 kW/m² K in the three effects.

Solution

For equal thermal loads in each effect, i.e. $Q_1 = Q_2 = Q_3$,

$$U_1 A_1 \Delta T_1 = U_2 A_2 \Delta T_2 = U_3 A_3 \Delta T_3$$

or, for equal areas in each effect,

$$U_1 \Delta T_1 = U_2 \Delta T_2 = U_3 \Delta T_3$$

In this case,

$$3{\cdot}1\Delta T_1 = 2{\cdot}3\Delta T_2 = 1{\cdot}1\Delta T_3$$

$$\therefore \qquad \Delta T_1 = 0{\cdot}742\Delta T_2$$

and

$$\Delta T_3 = 2{\cdot}091\Delta T_2$$

Now

$$\Sigma \Delta T = \Delta T_1 + \Delta T_2 + \Delta T_3 = (395 - 325) = 70\,\text{degK}$$

$$\therefore \qquad 0{\cdot}742\Delta T_2 + \Delta T_2 + 2{\cdot}091\Delta T_2 = 70\,\text{degK}$$

$$\therefore \qquad \Delta T_2 = 18{\cdot}3\,\text{degK}$$

and

$$\Delta T_1 = 13{\cdot}5\,\text{degK}, \quad \Delta T_3 = 38{\cdot}2\,\text{degK}$$

The temperatures in each effect are then:

$$T_1 = (395 - 13{\cdot}5) \quad = 381{\cdot}5\,\text{K}$$

$$T_2 = (381{\cdot}5 - 18{\cdot}3) = 363{\cdot}2\,\text{K}$$

$$\underline{T_3 = (363{\cdot}2 - 38{\cdot}2) = 325\,\text{K}}$$

Problem 14.14

A three-stage evaporator is fed with 1·25 kg/s of a liquor which is concentrated from 10% to 40% solids. The heat transfer coefficients may be taken as 3·1, 2·5, and 1·7 kW/m² K.

Calculate the steam at 170 kN/m² required per second and the temperature distribution in the three effects:

(a) if the feed is at 294 K;
(b) if the feed is at 355 K.

Forward feed is used in each instance, and the values of U may be considered the same for the two systems. The boiling-point in the third effect is 325 K; the liquor has no boiling-point rise.

Solution

(a) In the absence of data to the contrary, the specific heat will be taken as 4·18 kJ/kg K.

Making a mass balance:

	Solids (kg/s)	Liquor (kg/s)	Total (kg/s)
Feed	0·125	1·125	1·250
Product	0·125	0·188	0·313
Evaporation	—	0·937	0·937

∴ $$D_1 + D_2 + D_3 = 0.937\,\text{kg/s}$$

For steam at $170\,\text{kN/m}^2$, $$T_0 = 388\,\text{K}$$

∴ $$\Sigma\Delta T = (388 - 325) = 63\,\text{degK}$$

From equation 14.8:

$$3.1\Delta T_1 = 2.5\Delta T_2 = 1.7\Delta T_3$$

and hence $\Delta T_1 = 15.5\,\text{degK}$, $\Delta T_2 = 19\,\text{degK}$,   and $\Delta T_3 = 28.5\,\text{degK}$

Allowing for the cold feed (294 K), it will be assumed that:

$$\Delta T_1 = 20\,\text{degK},\quad \Delta T_2 = 17\,\text{degK},\quad \text{and}\quad \Delta T_3 = 26\,\text{degK}$$

Thus:
$$T_0 = 388\,\text{K},\quad \lambda_0 = 2216\,\text{kJ/kg}$$
$$T_1 = 368\,\text{K},\quad \lambda_1 = 2270\,\text{kJ/kg}$$
$$T_2 = 351\,\text{K},\quad \lambda_2 = 2312\,\text{kJ/kg}$$
$$T_3 = 325\,\text{K},\quad \lambda_3 = 2376\,\text{kJ/kg}$$

Making a heat balance across each effect,

(1) $2216D_0 = 1.25 \times 4.18(368 - 294) + 2270D_1$ or $D_0 = 0.175 + 1.024D_1$

(2) $2270D_1 + (1.25 - D_1)4.18(368 - 351) = 2312D_2$ or $D_2 = 0.951D_1 + 0.038$

(3) $2312D_2 + (1.25 - D_1 - D_2)4.18(351 - 325) = 2376D_3$

$$\text{or}\quad D_3 = 0.927D_2 - 0.046D_1 + 0.057$$

Noting that $D_1 + D_2 + D_3 = 0.937\,\text{kg/s}$, these equations may be solved to give:

$$D_0 = 0.472\,\text{kg/s},\quad D_1 = 0.290\,\text{kg/s},\quad D_2 = 0.313\,\text{kg/s},\quad D_3 = 0.334\,\text{kg/s}$$

The area of each effect is then:

$$A_1 = D_0\lambda_0/U_1\,\Delta T_1 = (0.472 \times 2216)/(3.1 \times 20) = 16.9\,\text{m}^2$$

$$A_2 = D_1\lambda_1/U_2\,\Delta T_2 = (0.290 \times 2270)/(2.5 \times 17) = 15.5\,\text{m}^2$$

$$A_3 = D_2\lambda_2/U_3\,\Delta T_3 = (0.334 \times 2312)/(1.7 \times 26) = 17.4\,\text{m}^2$$

These are probably sufficiently close for a first approximation and hence the steam consumption,

$$D_0 = 0.472 \text{ kg/s}$$

and the temperatures in each effect are

(1) 368 K,   (2) 351 K,   (3) 325 K

(b) In this case the feed is much hotter and hence the modification to the estimated values of ΔT will be less. Assume

$$\Delta T_1 = 17 \text{ degK}, \quad \Delta T_2 = 18 \text{ degK}, \quad \text{and} \quad \Delta T_3 = 28 \text{ degK}$$

Thus:
$$T_0 = 388 \text{ K}, \quad \lambda_0 = 2216 \text{ kJ/kg}$$
$$T_1 = 371 \text{ K}, \quad \lambda_1 = 2262 \text{ kJ/kg}$$
$$T_2 = 353 \text{ K}, \quad \lambda_2 = 2308 \text{ kJ/kg}$$
$$T_3 = 325 \text{ K}, \quad \lambda_3 = 2376 \text{ kJ/kg}$$

The heat balances are now:

(1) $2216 D_0 = 1.25 \times 4.18(371 - 355) + 2262 D_1$ or $D_0 = 0.038 + 1.021 D_1$

(2) $2262 D_1 + (1.25 - D_1) 4.18(371 - 353) = 2308 D_2$ or $D_2 = 0.948 D_1 + 0.041$

(3) $2308 D_2 + (1.25 - D_1 - D_2) 4.18(353 - 325) = 2376 D_3$

$$\text{or} \quad D_3 = 0.922 D_2 - 0.049 D_1 + 0.062$$

Again, $D_1 + D_2 + D_3 = 0.937 \text{ kg/s}$

and $D_0 = 0.331 \text{ kg/s}, \quad D_1 = 0.287 \text{ kg/s}, \quad D_2 = 0.313 \text{ kg/s}, \quad \text{and} \quad D_3 = 0.337 \text{ kg/s}$

$$\therefore \qquad A_1 = (0.331 \times 2216)/(3.1 \times 17) = 13.9 \text{ m}^2$$
$$A_2 = (0.287 \times 2262)/(2.5 \times 18) = 14.4 \text{ m}^2$$
$$A_3 = (0.313 \times 2308)/(1.7 \times 28) = 15.1 \text{ m}^2$$

These are close enough for design purposes and hence, the steam consumption,

$$D_0 = 0.331 \text{ kg/s}$$

and the temperatures in each effect are:

(1) 371 K,   (2) 353 K,   (3) 325 K

Problem 14.15

An evaporator operating on the thermo-recompression principle employs a steam ejector to maintain atmospheric pressure over the boiling liquid. This ejector uses 0.14 kg/s of steam at 650 kN/m² superheated 100 K and produces a pressure in the steam chest of 205 kN/m². A condenser removes surplus vapour from the atmospheric pressure line.

What is the capacity and economy of the system?
How could the economy be improved?

Data

Properties of the ejector:

Nozzle efficiency, 0·95.
Efficiency of momentum transfer, 0·80.
Efficiency of compression, 0·90.

The feed enters the evaporator at 295 K and thick liquor is withdrawn at a rate of 0·025 kg/s. The concentrated liquor exhibits a boiling-point rise of 10 degK. The plant is lagged sufficiently to render negligible the heat losses to the surroundings.

Solution

Let P_1 = pressure of live steam = 650 kN/m² and P_2 = pressure of entrained steam = 101·3 kN/m².

The enthalpy of the live steam at 650 kN/m² and (435 + 100) = 535 K, H_1 = 2970 kJ/kg.

Therefore H_2, the enthalpy after isentropic expansion from 650 to 101·3 kN/m², using an enthalpy–entropy chart, is H_2 = 2605 kJ/kg and the dryness fraction, x_2 = 0·97. The enthalpy of the steam after actual expansion to 101·3 kN/m² is given by H_2', where:

$$(H_1 - H_2') = 0{\cdot}95(2970 - 2605) = 347\,\text{kJ/kg}$$

and
$$H_2' = (2970 - 347) = 2623\,\text{kJ/kg}$$

At P_2 = 101·3 kN/m², $\lambda = 2258\,\text{kJ/kg}$

and the dryness after expansion but before entrainment x_2' is given by:

$$(x_2' - x_2)\lambda = (1 - e_1)(H_1 - H_2)$$

∴
$$(x_2' - 0{\cdot}97)\,2258 = (1 - 0{\cdot}95)(2970 - 2605)$$

∴
$$x_2' = 0{\cdot}978.$$

If x_2'' is the dryness after expansion *and* entrainment, then:

$$(x_2'' - x_2')\lambda = (1 - e_3)(H_1 - H_2')$$

∴
$$(x_2'' - 0{\cdot}978)\,2258 = (1 - 0{\cdot}80)(2970 - 2623)$$

$$x_2'' = 1{\cdot}00$$

Assuming that the steam at the discharge pressure P_3 = 205 kN/m² is also saturated, i.e. x_3 = 1·00, then from a chart H_3 the enthalpy of the mixture at the start of compression in the diffuser section at 101·3 kN/m² is H_3 = 2675 kJ/kg. Again assuming the entrained steam is also saturated, the enthalpy of the mixture after isentropic compression in the diffuser from 101·3 to 205 kN/m², H_4 = 2810 kJ/kg.

The entrainment ratio is then:

$$(m_2/m_1) = \{[(H_1 - H_2)/(H_4 - H_3)] \, e_1 \, e_2 \, e_3 - 1\}$$

where e_1, e_2 and e_3 are the efficiency of the nozzle, momentum transfer and compression, respectively.

$$\therefore \qquad (m_2/m_1) = \{[(2970 - 2605)/(2810 - 2675)] \, 0{\cdot}95 \times 0{\cdot}80 \times 0{\cdot}90 - 1\}$$

$$= 0{\cdot}85 \text{ kg vapour entrained/kg live steam}$$

It was assumed that $x_3 = 1{\cdot}0$. This may be checked by:

$$x_3 = [x_2 + x_4(m_2/m_1)]/(1 + m_2/m_1)$$

$$= (1{\cdot}0 + 1{\cdot}0 \times 0{\cdot}85)/(1 + 0{\cdot}85) = 1{\cdot}0$$

Thus with a flow of $0{\cdot}14$ kg/s live steam, the vapour entrained at $101{\cdot}3$ kN/m² is $(0{\cdot}14 \times 0{\cdot}85) = 0{\cdot}12$ kg/s, giving $0{\cdot}26$ kg/s steam saturated at 205 kN/m² to the calandria.

Allowing a 10 degK boiling-point rise, the temperature of boiling liquor in the unit is $T_1' = 383$ K and taking the specific heat as $4{\cdot}18$ kJ/kg K,

$$D_0 \lambda_0 = W C_p (T_1' - T_f) + D_1 \lambda_1$$

$$\therefore \qquad 0{\cdot}26 \times 2200 = W \times 4{\cdot}18 (393 - 295) + D_1 \times 2258$$

$$572 = 368W + 2258 D_1$$

But

$$W - D_1 = 0{\cdot}025 \text{ kg/s}$$

$$\therefore \qquad D_1 = 0{\cdot}214 \text{ kg/s}$$

$$\therefore \qquad \text{economy of system} = 0{\cdot}214/0{\cdot}14 = \underline{1{\cdot}53}$$

The capacity, in terms of the throughput of solution, is:

$$W = 0{\cdot}214 + 0{\cdot}025$$

$$= \underline{0{\cdot}239 \text{ kg/s}}$$

Apart from increasing the efficiency of the ejector, the economy of the system might be increased by operating with a higher live-steam pressure, increasing the pressure in the vapour space, and by using the vapour not returned to the ejector to preheat the feed solution.

Problem 14.16

A single-effect evaporator is used to concentrate $0{\cdot}075$ kg/s of a 10% caustic soda liquor to 30%. The unit employs forced circulation in which the liquor is pumped through the vertical tubes of the calandria which are 32 mm o.d. by 28 mm i.d. and $1{\cdot}2$ m long.

Steam is supplied at 394 K dry and saturated, and the boiling-point rise of the 30% solution is 15 degK. If the overall heat transfer coefficient is $1{\cdot}75$ kW/m² K, how many tubes do you suggest should be used and what material of construction would you advise for the evaporator?

The latent heat of vaporisation under these conditions is 2270 kJ/kg.

Solution

Making a mass balance:

	Solids (kg/s)	Liquor (kg/s)	Total (kg/s)
Feed	0·0075	0·0675	0·0750
Product	0·0075	0·0175	0·0250
Evaporation	—	0·0500	0·0500

The temperature of boiling liquor in the tubes, assuming atmospheric pressure, $T_1' = 373 + 15 = 388$ K. In the absence of any other data it will be assumed that the solution enters at 373 K and the specific heat will be taken as 4·18 kJ/kg K.

A heat balance then gives:

$$D_0 \lambda_0 = WC_p(T_1' - T_f) + D_1 \lambda_1$$

$$= 0·0750 \times 4·18(388 - 373) + 0·050 \times 2270$$

$$= 118·2 \, \text{kW}$$

$$\Delta T_1 = (394 - 388) = 6 \, \text{degK}$$

and the area,
$$A_1 = D_0 \lambda_0 / U_1 \Delta T_1$$

$$= 118·2/(1·75 \times 6) = 11·25 \, \text{m}^2$$

The tube o.d. is 0·032 m and the outside area per unit length of tubing,

$$\pi \times 0·032 = 0·101 \, \text{m}^2/\text{m}$$

∴
$$\text{total length of tubing required} = 11·25/0·101 = 112 \, \text{m}$$

$$\text{and number of tubes required} = 112/1·2 = \underline{\underline{93}}$$

Mild steel cannot cope with caustic soda solutions, and stainless steel has limitations at higher temperatures. Aluminium bronze, copper, nickel, and nickel–copper alloys can be used, together with neoprene and butyl rubber, though from the cost viewpoint and the need for a good conductivity, graphite tubes would probably be specified.

Problem 14.17

A steam-jet booster compresses 0·1 kg/s of dry and saturated vapour from 3·4 kN/m² to 13·4 kN/m². High-pressure steam consumption is 0·05 kg/s at 690 kN/m².
(a) What must be the condition of the HP steam for the booster discharge to be superheated 20 degK?
(b) What is the overall efficiency of the booster if the compression efficiency is assumed equal to 1?

Solution

(a) Considering the outlet stream, at 13·4 kN/m² and 20 degK superheat, this has an enthalpy, $H_4 = 2638$ kJ/kg.

The enthalpy of the entrained vapours, $H_3' = 2540$ kJ/kg assuming saturation.

If H_1 is the enthalpy of the HP steam, then an enthalpy balance gives:

$$0.05H_1 + 0.1 \times 2540 = 0.15 \times 2638$$

∴
$$H_1 = 2834 \text{ kJ/kg}$$

At 690 kN/m², this corresponds to a temperature of 453 K.

At 690 kN/m², steam is saturated at 438 K, and the HP steam must be supplied with 15 degK superheat.

(b) Assuming the HP is expanded isentropically from 690 kN/m² (and 453 K) to 3·4 kN/m², H_2 (taking an efficiency of unity) = 2045 kJ/kg.

If the outlet stream is attained by isentropic compression from 3·4 kN/m² to 13·4 kN/m², then $H_3 = 2435$ kJ/kg and the efficiency is given by:

$$(0.1/0.05) = [(2834 - 2045)/(2638 - 2435)]\,\eta - 1$$

∴
$$\eta = 0.77$$

Problem 14.18

A triple-effect backward-feed evaporator concentrates 5 kg/s of liquor from 10% to 50% solids. Steam at 375 kN/m² is available and the condenser operates at 13·5 kN/m². Find the area required in each effect (areas to be equal) and the economy of the unit.

Assume that the specific heat is 4·18 kJ/kg K at all concentrations, and that there is no boiling-point rise. The overall heat transfer coefficients are 2·3, 2·0, and 1·7 kW/m² K respectively in the three effects. The feed enters the third effect at 300 K.

Solution

As a variation, this problem will be solved using Storrow's Method (Vol. 2, Section 14.3.3). A mass balance gives the total evaporation as follows:

	Solids (kg/s)	Liquor (kg/s)	Total (kg/s)
Feed	0·50	4·50	5·0
Product	0·50	0·50	1·0
Evaporation	—	4·0	4·0

For steam at 375 kN/m², $T_0 = 414$ K

For steam at 13·5 kN/m², $T_3 = 325$ K

∴ $\Sigma \Delta T = (414 - 325) = 89$ degK

From equation 14.8:
$$2\cdot3\Delta T_1 = 2\cdot0\Delta T_2 = 1\cdot7\Delta T_3$$

and hence: $\quad \Delta T_1 = 26\,\text{degK}, \quad \Delta T_2 = 29\,\text{degK}, \quad \Delta T_3 = 34\,\text{degK}$

Modifying the values to take account of the feed temperature, it will be assumed that:

$$\Delta T_1 = 27\,\text{degK}, \quad \Delta T_2 = 30\,\text{degK}, \quad \text{and} \quad \Delta T_3 = 32\,\text{degK}$$

and hence $\quad T_1 = 387\,\text{K}, \quad T_2 = 357\,\text{K}, \quad \text{and} \quad T_3 = 325\,\text{K}$

As a first approximation, assume equal evaporation in each effect, i.e.

$$D_1 = D_2 = D_3 = 1\cdot33\,\text{kg/s}$$

With backward feed, the liquor has to be raised to its boiling-point as it enters each effect.

Heat required to raise the feed to the second effect to its boiling-point

$$= (4\cdot0 - 1\cdot33)4\cdot18(357 - 325) = 357\cdot1\,\text{kW}$$

Heat required to raise the feed to the first effect to its boiling-point

$$= (4\cdot0 - 2 \times 1\cdot33)4\cdot18(387 - 357) = 168\cdot1\,\text{kW}$$

At $T_0 = 414\,\text{K}$, $\qquad\qquad \lambda_0 = 2140\,\text{kJ/kg}$

At $T_3 = 325\,\text{K}$, $\qquad\qquad \lambda_3 = 2376\,\text{kJ/kg}$

a mean value of 2258 kJ/kg which will be assumed to be the value of the latent heat in all effects.

The relation between the heat transferred in each effect and in the condenser is given by:

$$Q_1 - 168\cdot1 = Q_2 = Q_3 + 357\cdot1 = Q_c + 357\cdot1 + [5\cdot0 \times 4\cdot18(325 - 300)]$$
$$= Q_c + 879\cdot6$$

The total evaporation $\qquad (Q_2 + Q_3 + Q_c)/2258 = 4\cdot0$

or $\qquad\qquad Q_2 + (Q_2 - 357\cdot1) + (Q_2 - 879\cdot6) = 9032$

$\therefore$ $\qquad\qquad\qquad Q_2 = 3423 = A\Delta T_2 \times 2\cdot0$

$\qquad\qquad\qquad Q_3 = 3066 = A\Delta T_3 \times 1\cdot7$

$\qquad\qquad\qquad Q_1 = 3591 = A\Delta T_1 \times 2\cdot3$

$\therefore$ $\quad A\Delta T_1 = 1561\,\text{m}^2\,\text{K}, \quad A\Delta T_2 = 1712\,\text{m}^2\,\text{K}, \quad \text{and} \quad A\Delta T_3 = 1804\,\text{m}^2\,\text{K}$

$$\Delta T_1 + \Delta T_2 + \Delta T_3 = 89\,\text{degK}$$

Values of ΔT_1, ΔT_2, and ΔT_3 are now chosen by trial and error to give equal values of A in each effect as follows:

ΔT_1	A_1	ΔT_2	A_2	ΔT_3	A_3
27	57·8	30	57·1	32	56·4
27·5	56·8	30·5	56·1	31	58·1
27·25	57·3	30·25	56·6	31·5	57·3

These areas are now approximately equal and the assumed values of ΔT are acceptable.

$$\text{Economy} = 4{\cdot}0/(3591/2258) = \underline{\underline{2{\cdot}52}}$$

and the area to be specified for each effect = $\underline{\underline{57\,\text{m}^2}}$.

Problem 14.19

A double-effect evaporator of the climbing film type is connected so that the feed passes through two preheaters—one heated by vapour from the first effect and the other by vapour from the second effect. The condensate from the first effect is passed into the steam space of the second. The temperature of the feed is initially 289 K, after the first heater 348 K and after the second heater 383 K. The vapour temperature in the first effect is 398 K and in the second 373 K. The feed is 0·25 kg/s and the steam is dry and saturated at 413 K.

Find the economy of the unit if the evaporation rate is 0·125 kg/s.

Solution

A heat balance across the first effect gives:

$$D_0 \lambda_0 = WC_p(T_1 - T_f) + D_1 \lambda_1$$

where T_f is the temperature of the feed leaving the second preheater, i.e. 383 K.

When $T_0 = 413$ K, $\lambda_0 = 2140\,\text{kJ/kg}$

When $T_1 = 398$ K, $\lambda_1 = 2190\,\text{kJ/kg}$

Taking $C_p = 4{\cdot}18\,\text{kJ/kg K}$ throughout,

$$2140 D_0 = 0{\cdot}25 \times 4{\cdot}18(398 - 383) + 2190 D_1$$

$$\therefore \qquad D_0 = 0{\cdot}0073 + 0{\cdot}023 D_1 \text{ kg/s}$$

Assuming that the first preheater is heated by steam from the first effect, then the heat transferred in this unit

$$= 0{\cdot}25 \times 4{\cdot}18(348 - 289) = 61{\cdot}7\,\text{kW}$$

and hence the steam condensed in the preheater = $(61{\cdot}7/2190) = 0{\cdot}028$ kg/s.

Therefore vapour from the first effect which is condensed in the steam space of the second effect = $(D_1 - 0{\cdot}028)$ kg/s.

A heat balance on the second effect is thus:

$$(D_1 - 0{\cdot}028)\lambda_1 + (W - D_1)C_p(T_1 - T_2) = D_2 \lambda_2$$

Since the total evaporation = 0·125 kg/s,

$$D_2 = 0{\cdot}125 - D_1$$

and $(D_1 - 0{\cdot}028)2190 + (0{\cdot}25 - D_1)4{\cdot}18(398 - 373) = (0{\cdot}125 - D_1)2256$

$$\therefore \qquad D_1 = 0{\cdot}073\,\text{kg/s}$$

$$\therefore \qquad D_0 = 0\cdot0073 + 1\cdot023 \times 0\cdot073$$

$$= 0\cdot082\,\text{kg/s}$$

and the economy
$$= (0\cdot125/0\cdot082) = \underline{\underline{1\cdot5\,\text{kg/kg}}}$$

Problem 14.20

A triple-effect evaporator is fed with 5 kg/s of a liquor containing 15% solids. The concentration in the last effect, which operates at 13·5 kN/m², is 60% solids.

If the overall heat transfer coefficients are 2·5, 2·0, and 1·1 kW/m² K, respectively, and the steam is fed at 388 K to the first effect, determine the temperature distribution and the area of heating surface required in each effect. The calandrias are to be identical.

What is the economy and what is the heat load on the condenser?

Feed temperature = 294 K.

Specific heat of all liquors = 4·18 kJ/kg K.

If the unit is run as a backward-feed system, the coefficients are 2·3, 2·0, and 1·6 kW/m² K. Under these conditions determine the new temperatures, heat economy, and heating surface required.

Solution

(a) *Forward feed*

A mass balance gives:

	Solids (kg/s)	Liquor (kg/s)	Total (kg/s)
Feed	0·75	4·25	5·0
Product	0·75	0·50	1·25
Evaporation	—	3·75	3·75

For steam saturated at 13·5 kN/m²,

$$T_3 = 325\,\text{K} \quad \text{and} \quad \lambda_3 = 2375\,\text{kJ/kg}$$

$$T_0 = 388\,\text{K} \quad \text{and} \quad \lambda_0 = 2216\,\text{kJ/kg}$$

$$\therefore \qquad \Sigma\Delta T = (388 - 325) = 63\,\text{degK}$$

From equation 14.8:

$$U_1\,\Delta T_1 = U_2\,\Delta T_2 = U_3\,\Delta T_3$$

$$\therefore \qquad 2\cdot5\Delta T_1 = 2\cdot0\Delta T_2 = 1\cdot1\Delta T_3$$

$$\therefore \qquad \Delta T_1 = 14\,\text{degK}, \quad \Delta T_2 = 17\cdot5\,\text{degK}, \quad \text{and} \quad \Delta T_3 = 31\cdot5\,\text{degK}$$

Allowing for the cold feed it will be assumed that:

$$\Delta T_1 = 18 \text{ degK}, \quad \Delta T_2 = 16 \text{ degK}, \quad \text{and} \quad \Delta T_3 = 29 \text{ degK}$$

$$\therefore \qquad T_0 = 388 \text{ K}, \quad \lambda_0 = 2216 \text{ kJ/kg}$$

$$T_1 = 370 \text{ K}, \quad \lambda_1 = 2266 \text{ kJ/kg}$$

$$T_2 = 354 \text{ K}, \quad \lambda_2 = 2305 \text{ kJ/kg}$$

$$T_3 = 325 \text{ K}, \quad \lambda_3 = 2375 \text{ kJ/kg}$$

Making heat balances over each effect:

(1) $D_0 \lambda_0 = W C_p (T_1 - T_f) + D_1 \lambda_1$ or $2216 D_0 = 5.0 \times 4.18 (370 - 294) + 2266 D_1$

(2) $D_1 \lambda_1 + (W - D_1) C_p (T_1 - T_2) = D_2 \lambda_2$

$$\text{or} \quad 2266 D_1 + (5.0 - D_1) 4.18 (370 - 354) = 2305 D_2$$

(3) $D_2 \lambda_2 + (W - D_1 - D_2) C_p (T_2 - T_3) = D_3 \lambda_3$

$$\text{or} \quad 2305 D_2 + (5.0 - D_1 - D_2) 4.18 (354 - 325) = 2375 D_3$$

Also $(D_1 + D_2 + D_3) = 3.75 \text{ kg/s}$

and solving simultaneously:

$$D_0 = 1.90 \text{ kg/s}, \quad D_1 = 1.16 \text{ kg/s}, \quad D_2 = 1.25 \text{ kg/s}, \quad \text{and} \quad D_3 = 1.35 \text{ kg/s}$$

The areas are now given by:

$$A_1 = D_0 \lambda_0 / U_1 \Delta T_1 = (1.90 \times 2216)/(2.5 \times 18) = 93.6 \text{ m}^2$$

$$A_2 = D_1 \lambda_1 / U_2 \Delta T_2 = (1.16 \times 2266)/(2.0 \times 16) = 82.1 \text{ m}^2$$

$$A_3 = D_2 \lambda_2 / U_3 \Delta T_3 = (1.25 \times 2305)/(1.1 \times 29) = 90.3 \text{ m}^2$$

It is fairly obvious that the modification for the cold feed has been incorrect and it will now be assumed that:

$$\Delta T_1 = 19 \text{ degK}, \quad \Delta T_2 = 15 \text{ degK}, \quad \text{and} \quad \Delta T_3 = 29 \text{ degK}$$

$$\therefore \qquad T_0 = 388 \text{ K} \quad \text{and} \quad \lambda_0 = 2216 \text{ kJ/kg}$$

$$T_1 = 369 \text{ K} \quad \text{and} \quad \lambda_1 = 2267 \text{ kJ/kg}$$

$$T_2 = 354 \text{ K} \quad \text{and} \quad \lambda_2 = 2305 \text{ kJ/kg}$$

$$T_3 = 325 \text{ K} \quad \text{and} \quad \lambda_3 = 2375 \text{ kJ/kg}$$

The heat balances now become:

(1) $2216 D_0 = (5.0 \times 4.18)(369 - 294) + 2267 D_1$

(2) $2267 D_1 + (5.0 - D_1) 4.18 (369 - 354) = 2305 D_2$

(3) $2305 D_2 + (5.0 - D_1 - D_2) 4.18 (354 - 325) = 2375 D_3$

and solving:

$$D_0 = 1.89 \text{ kg/s}, \quad D_1 = 1.16 \text{ kg/s}, \quad D_2 = 1.25 \text{ kg/s}, \quad \text{and} \quad D_3 = 1.34 \text{ kg/s}$$

Thence:
$$A_1 = (1 \cdot 89 \times 2216)/(2 \cdot 5 \times 19) = 88 \cdot 2 \, \text{m}^2$$
$$A_2 = (1 \cdot 16 \times 2267)/(2 \cdot 0 \times 15) = 87 \cdot 8 \, \text{m}^2$$
$$A_3 = (1 \cdot 25 \times 2305)/(1 \cdot 1 \times 29) = 90 \cdot 3 \, \text{m}^2$$

which are much closer.

Thus, the temperature distribution is:

$$\underline{(1) \; 369 \, \text{K}, \quad (2) \; 354 \, \text{K}, \quad (3) \; 325 \, \text{K}}$$

The area in each effect should be $\underline{89 \, \text{m}^2}$.

The economy is given as $(3 \cdot 75/1 \cdot 89) = \underline{\underline{2 \cdot 0}}$ and the heat load on the condenser,

$$D_3 \lambda_3 = (1 \cdot 34 \times 2375)$$
$$= \underline{\underline{31 \cdot 83 \, \text{kW}}}$$

(b) *Backward feed*

As before,
$$D_1 + D_2 + D_3 = 3 \cdot 75 \, \text{kg/s}$$

and
$$\Sigma \Delta T = (388 - 325) = 63 \, \text{degK}$$

In this case,
$$2 \cdot 3 \Delta T_1 = 2 \cdot 0 \Delta T_2 = 1 \cdot 6 \Delta T_3$$

$\therefore$
$$\Delta T_1 = 17 \cdot 5 \, \text{degK}, \quad \Delta T_2 = 20 \cdot 0 \, \text{degK}, \quad \text{and} \quad \Delta T_3 = 25 \cdot 5 \, \text{degK}$$

Modifying for the cold feed, it will be assumed that:

$$\Delta T_1 = 18 \cdot 5 \, \text{degK}, \quad \Delta T_2 = 20 \cdot 5 \, \text{degK}, \quad \text{and} \quad \Delta T_3 = 24 \, \text{degK}$$

$\therefore$
$$T_0 = 388 \, \text{K}, \quad \lambda_0 = 2216 \, \text{kJ/kg}$$
$$T_1 = 369 \cdot 5 \, \text{K}, \quad \lambda_1 = 2266 \, \text{kJ/kg}$$
$$T_2 = 349 \, \text{K}, \quad \lambda_2 = 2318 \, \text{kJ/kg}$$
$$T_3 = 325 \, \text{K}, \quad \lambda_3 = 2375 \, \text{kJ/kg}$$

The heat balances become:

(1) $D_0 \lambda_0 = (W - D_3 - D_2) C_p (T_1 - T_2) + D_1 \lambda_1$

$\qquad\qquad\qquad$ or $\quad 2216 D_0 = (5 \cdot 0 - D_3 - D_2) 4 \cdot 18 (369 \cdot 5 - 349) + 2266 D_1$

(2) $D_1 \lambda_1 = (W - D_3) C_p (T_2 - T_3) + D_2 \lambda_2$

$\qquad\qquad\qquad$ or $\quad 2266 D_1 = (5 \cdot 0 - D_3) 4 \cdot 18 (349 - 325) + 2318 D_2$

(3) $D_2 \lambda_2 = W C_p (T_3 - T_f) + D_3 \lambda_3$ or $\quad 2318 D_2 = 5 \cdot 0 \times 4 \cdot 18 (325 - 294) + 2375 D_3$

Solving:

$$D_0 = 1 \cdot 62 \, \text{kg/s}, \quad D_1 = 1 \cdot 49 \, \text{kg/s}, \quad D_2 = 1 \cdot 28 \, \text{kg/s}, \quad \text{and} \quad D_3 = 0 \cdot 98 \, \text{kg/s}$$

The areas are:

$$A_1 = (1 \cdot 62 \times 2216)/(2 \cdot 3 \times 18 \cdot 5) = 84 \cdot 4 \, \text{m}^2$$

$$A_2 = (1 \cdot 49 \times 2266)/(2 \cdot 0 \times 20 \cdot 5) = 82 \cdot 4 \, \text{m}^2$$

$$A_3 = (1 \cdot 28 \times 2318)/(1 \cdot 6 \times 24) = 77 \cdot 3 \, \text{m}^2$$

which are probably close enough for design purposes.
Thus, the temperatures in each effect become:

$$(1) \ 369 \cdot 5 \, \text{K}, \quad (2) \ 349 \, \text{K}, \quad \text{and} \quad (3) \ 325 \, \text{K}$$

The economy is $(3 \cdot 75/1 \cdot 62) = \underline{\underline{2 \cdot 3}}$, and the area of each effect $= \underline{\underline{81 \, \text{m}^2}}$.

Problem 14.21

A double-effect forward-feed evaporator is required to give a product which contains 50% by weight of solids. Each effect has $10 \, \text{m}^2$ of heating surface and the heat transfer coefficients are known to be 2·8 and 1·7 kW/m² K in the first and second effects respectively. Dry and saturated steam is available at 375 kN/m² and the condenser operates at 13·5 kN/m². The concentrated solution exhibits a boiling-point rise of 3 degK.

What is the maximum feed rate if the feed contains 10% solids and is at a temperature of 310 K?

Assume a latent heat of 2330 kJ/kg and a specific heat of 4·18 kJ/kg under all conditions.

Solution

Making a mass balance on the basis of W kg/s feed:

	Solids (kg/s)	Liquor (kg/s)	Total (kg/s)
Feed	$0 \cdot 10W$	$0 \cdot 90W$	W
Product	$0 \cdot 10W$	$0 \cdot 10W$	$0 \cdot 20W$
Evaporation	—	$0 \cdot 80W$	$0 \cdot 80W$

Thus

$$D_1 + D_2 = 0 \cdot 8W \text{ kg/s}$$

At 375 kN/m²,

$$T_0 = 413 \text{ K}$$

At 13·5 kN/m²,

$$T_2 = 325 \text{ K}$$

and the temperature of the boiling liquor in the second effect,

$$T_2' = 328 \text{ K}$$

$$\therefore \qquad \Sigma \Delta T = (413 - 328) = 85 \text{ degK}$$

At this stage, values could be assumed for ΔT_1 and ΔT_2 and heat balances made until equal areas resulted. There are sufficient data, however, to enable an exact solution to be obtained as follows. Making heat balances:

1st effect

$$D_0 \lambda_0 = W(T_1 - T_f) + D_1 \lambda_1$$

or $\qquad AU_1(T_0 - T_1) = W(T_1 - T_f) + AU_2(T_1 - T_2')$

$\therefore \qquad 10 \times 2{\cdot}8(413 - T_1) = W \times 4{\cdot}18(T_1 - 310) + 10 \times 1{\cdot}7(T_1 - T_2')$

$\therefore \qquad W = (17{,}133 - 45T_1)/(4{\cdot}18T_1 - 1296)$ \hfill (i)

2nd effect

$$D_1 \lambda_1 = (W - D_1)(T_1 - T_2') + D_2 \lambda_2$$

But $\qquad D_1 = 10 \times 1{\cdot}7(T_1 - 328)/2330 = 0{\cdot}0073T_1 - 2393$

and $\qquad D_2 = 0{\cdot}8W - 0{\cdot}0073T_1 - 2{\cdot}393$

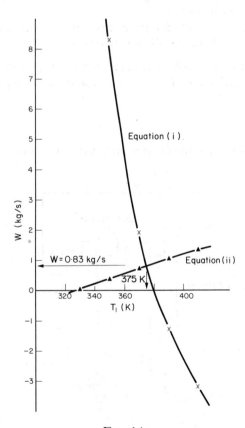

FIG. 14a

$$\therefore \quad (0{\cdot}0073T_1 - 2{\cdot}393)\,2330$$

$$= (W - 0{\cdot}0073T_1 + 2{\cdot}393)\,4{\cdot}18\,(T_1 - 328) + (0{\cdot}8W - 0{\cdot}0073T_1 - 2{\cdot}393)\,2330$$

$$\therefore \qquad\qquad W = (14T_1 - 7872 + 0{\cdot}0305T_1^2)/(4{\cdot}18T_1 + 493) \qquad\qquad \text{(ii)}$$

Equations (i) and (ii) are plotted in Fig. 14a and the two curves coincide when $T_1 = 375\,\text{K}$ and $W = 0{\cdot}83\,\text{kg/s}$.

Problem 14.22

You are required to consider proposals for concentrating fruit juice by evaporation. It is proposed to use a falling film evaporator and to incorporate a heat pump cycle with ammonia as the medium. The ammonia in vapour form enters the evaporator at 312 K and the water is evaporated from the juices at 287 K. The ammonia in the vapour–liquid mixture enters the condenser at 278 K and the vapour then passes to the compressor. It is estimated that the work in compressing the ammonia will be 150 kJ/kg of ammonia and that 2·28 kg of ammonia is cycled per kilogram of water evaporated. The following proposals are made for driving the compressor:

(a) to use a diesel engine drive taking 0·4 kg of fuel per MJ, the calorific value being 42 MJ/kg and the cost £0·02/kg;
(b) to pass steam, costing £0·01 for 10 kg through a turbine which operates at 70% isentropic efficiency, between 700 and 101·3 kN/m².

Explain by means of a diagram how this plant will work, including all necessary major items of equipment. Which method would you suggest for driving the compressor?

Solution

A simplified flow diagram of the plant is included as Fig. 14b.

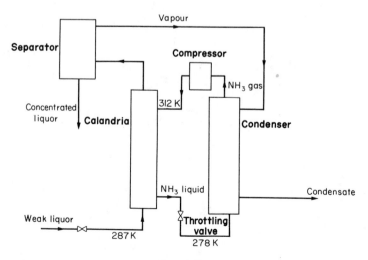

FIG. 14b

Considering the ammonia cycle

Ammonia gas will leave the condenser, probably saturated at low pressure, and enter the compressor which it leaves at high pressure and 312 K. In the calandria heat will be transferred to the liquor and the ammonia gas will be cooled to saturation, condense, and indeed may possibly leave the unit at 278 K as slightly sub-cooled liquid though still at high pressure. This liquid is then allowed to expand adiabatically in the throttling valve to the lower pressure during which some vaporisation will occur and the vapour–liquid mixture will enter the condenser with a dryness fraction of, say, 0·1–0·2. In the condenser heat is transferred from the condensing vapours and the liquid ammonia will leave the condenser probably just saturated though still at the low pressure. The cycle is then repeated.

Considering the liquor stream

Weak liquor will enter the plant and pass to the calandria where it is drawn up as a thin film by the partial vacuum caused by ultimate condensation of vapour in the condenser. Vaporisation takes place due to heat transfer from condensing ammonia in the calandria, and the vapour and concentrated liquor are then separated in a separator from which the concentrated liquor is drawn off as product. The vapours pass to the condenser where they are condensed by heat transfer to the evaporating ammonia and leave the plant as condensate. A final point is that any excess heat introduced by the compressor must be removed from the ammonia by means of a cooler.

Fuller details of the cycle and salient features of operation are given in Vol. 2, Section 14.4.2.

Choice of compressor drive (basis 1 kg water evaporated)

(a) *Diesel engine*

For 1 kg evaporation, ammonia circulated $= 2\cdot28$ kg and the work done in compressing the ammonia $= (150 \times 2\cdot28)$

$$= 342\,kJ \quad \text{or} \quad 0\cdot342\,MJ/kg \text{ evaporation}$$

For an output of 1 MJ, the engine consumes 0·4 kg fuel.

$$\therefore \qquad \text{fuel consumption} = (0\cdot4 \times 0\cdot342) \; = 0\cdot137\,kg/kg \text{ water evaporated}$$

$$\text{cost} = (0\cdot02 \times 0\cdot137) = 0\cdot002\,74\,£/kg \text{ water evaporated}$$

(b) *Turbine*

The work required is 0·342 MJ/kg evaporation.
Therefore with an efficiency of 70%, energy required from steam $= (0\cdot342 \times 100/70) = 0\cdot489$ MJ/kg.
Enthalpy of steam saturated at 700 kN/m^2 = 2764 kJ/kg.
Enthalpy of steam saturated at 101·3 kN/m^2 = 2676 kJ/kg.

$$\therefore \qquad \text{energy from steam} = (2764 - 2676) = 88\,kJ/kg \quad \text{or} \quad 0\cdot088\,MJ/kg$$

$$\therefore \qquad \text{steam required} = 0\cdot489/0\cdot088 = 5\cdot56\,kg/kg \text{ evaporation}$$

at a cost of $(0.01 \times 5.56/10) = \underline{\underline{0.0056 \text{ £/kg water evaporated}}}$

and hence the *diesel engine would be used for driving the compressor.*

Problem 14.23

A double-effect forward-feed evaporator is required to give a product consisting of 30% crystals and a mother liquor containing 40% by weight of dissolved solids. Heat transfer coefficients are 2·8 and 1·7 kW/m² K in the first and second effects respectively. Dry saturated steam is supplied at 375 kN/m² and the condenser operates at 13·5 kN/m².

(a) What area of heating surface is required in each effect (both effects to be identical) if the feed rate is 0·6 kg/s of liquor containing 20% by weight of dissolved solids and the feed temperature is 313 K?

(b) What is the pressure above the boiling liquid in the first effect? Take the specific heat capacity of the liquid as constant at 4·18 kJ/kg K and neglect effects of boiling-point rise and of hydrostatic head.

Solution

The final product contains 30% crystals and hence 70% solution containing 40% dissolved solids. The total percentage of dissolved and undissolved solids in the final product is $(0.30 + 0.40 \times 0.70) = 0.58$ or 58%, and the mass balance becomes:

	Solids (kg/s)	Liquor (kg/s)	Total (kg/s)
Feed	0·12	0·48	0·60
Product	0·12	0·087	0·207
Evaporation	—	$(D_1 + D_2) = 0.393$	0·393

At 375 kN/m², $T_0 = 413 \text{ K}$

and at 13·5 kN/m², $T_2 = 325 \text{ K}$

∴ $\Sigma \Delta T = (413 - 325) = 88 \text{ degK}$

From equation 14.8: $2.8 \Delta T_1 = 1.7 \Delta T_2$

∴ $\Delta T_1 = 33 \text{ degK}$ and $\Delta T_2 = 55 \text{ degK}$

Allowing for the cold feed, it will be assumed that:

$\Delta T_1 = 40 \text{ degK}$ and $\Delta T_2 = 48 \text{ degK}$

Thus: $T_0 = 413 \text{ K}, \quad \lambda_0 = 2140 \text{ kJ/kg}$

$T_1 = 373 \text{ K}, \quad \lambda_1 = 2257 \text{ kJ/kg}$

$T_2 = 325 \text{ K}, \quad \lambda_1 = 2375 \text{ kJ/kg}$

Taking the specific heat of the liquor as 4·18 kJ/kg K at all times, then the heat balance over each effect becomes:

(1) $D_0 \lambda_0 = WC_p(T_1 - T_f) + D_1 \lambda_1$ or $2140D_0 = 0.6 \times 4.18(373 - 313) + 2257D_1$

(2) $D_1 \lambda_1 + (W - D_1) C_p(T_1 - T_2) = D_2 \lambda_2$

$$\text{or}\quad 2257D_1 + (0.6 - D_1)4.18(373 - 325) = 2375D_2$$

Solving: $D_0 = 0.263\,\text{kg/s},\quad D_1 = 0.184\,\text{kg/s},\quad \text{and}\quad D_2 = 0.209\,\text{kg/s}$

The areas are now given by:

$$A_1 = D_0 \lambda_0 / U_1 \Delta T_1 = (0.263 \times 2140)/(2.8 \times 40) = 5.04\,\text{m}^2$$
$$A_2 = D_1 \lambda_1 / U_2 \Delta T_2 = (0.184 \times 2257)/(1.7 \times 48) = 5.08\,\text{m}^2$$

which show excellent agreement.

∴   the area of heating to be specified $= \underline{5.1\,\text{m}^2}$ in each effect

The temperature of liquor boiling in the first effect, assuming no boiling-point rise, $=$ 373 K at which pressure steam is saturated at 101·3 kN/m². The pressure in the first effect is, therefore, <u>atmospheric</u>.

Problem 14.24

1·9 kg/s of a liquid containing 10% by weight of dissolved solids is fed at 338 K to a forward-feed double-effect evaporator. The product consists of 25% by weight of solids and a mother liquor containing 25% by weight of dissolved solids. The steam fed to the first effect is dry and saturated at 240 kN/m² and the pressure in the second effect of this evaporator is 20 kN/m². The specific heat of the solid can be taken as 2·5 kJ/kg K whether it is in solid form or in solution and the heat of solution may be neglected. The mother liquor exhibits a boiling-point rise of 6 degK. If the two effects are identical, what area is required if the heat transfer coefficients in the first and second effects are 1·7 and 1·1 kW/m² K respectively?

Solution

The percentage by mass of dissolved and undissolved solids in the final product $=$ $0.25 \times 75 + 25 = 43.8\%$ and hence the mass balance becomes:

	Solids (kg/s)	Liquor (kg/s)	Total (kg/s)
Feed	0·19	1·71	1·90
Product	0·19	0·244	0·434
Evaporation	—	$(D_1 + D_2) = 1.466$	1·466

At 240 kN/m², $T_0 = 399\,\text{K}$

At 20 kN/m², $T_2 = 333\,\text{K}$

$$\therefore \qquad\qquad T_2' = 333 + 6 = 339 \, \text{K}$$

and, allowing for the boiling-point rise in the first effect,

$$\Delta T_1 + \Delta T_2 = (399 - 339) - 6 = 54 \, \text{degK}$$

From equation 14.8: $\qquad\qquad 1 \cdot 7 \Delta T_1 = 1 \cdot 1 \Delta T_2$

$$\therefore \qquad\qquad \Delta T_1 = 21 \, \text{degK} \quad \text{and} \quad \Delta T_2 = 33 \, \text{degK}$$

Modifying these values to allow for the cold feed, it will be assumed that:

$$\Delta T_1 = 23 \, \text{degK} \quad \text{and} \quad \Delta T_2 = 31 \, \text{degK}$$

Assuming that the liquor exhibits a 6 degK boiling-point rise at all concentrations, then, with T_1' as the temperature of boiling liquor in the first effect and T_2' that in the second effect:

$$T_0 = 399 \, \text{K} \quad \text{at which} \quad \lambda_0 = 2185 \, \text{kJ/kg}$$

$$T_1' = (399 - 23) = 376 \, \text{K}$$

$$T_1 = (376 - 6) = 370 \, \text{K} \quad \text{at which} \quad \lambda_1 = 2266 \, \text{kJ/kg}$$

$$T_2' = 339 \, \text{K}$$

$$T_2 = 333 \, \text{K} \quad \text{at which} \quad \lambda_2 = 2258 \, \text{K}$$

Making a heat balance over each effect:

(1) $D_0 \lambda_0 = W C_p (T_1' - T_f) + D_1 \lambda_1 \quad$ or $\quad 2185 D_0 = 1 \cdot 90 \times 2 \cdot 5(376 - 338) + 2266 D_1$

(2) $D_1 \lambda_1 + (W - D_1) C_p (T_1' - T_2') = D_2 \lambda_2$

$$\text{or} \quad 2358 D_2 = (1 \cdot 90 - D_1) 2 \cdot 5(376 - 339) + 2266 D_1$$

Solving, $\quad D_0 = 0 \cdot 833 \, \text{kg/s}, \quad D_1 = 0 \cdot 724 \, \text{kg/s}, \quad \text{and} \quad D_2 = 0 \cdot 742 \, \text{kg/s}$

The areas are then given by:

$$A_1 = D_0 \lambda_0 / U_1 (T_0 - T_1') = 0 \cdot 833 \times 2185 / [1 \cdot 7(399 - 376)] = 46 \cdot 7 \, \text{m}^2$$

$$A_2 = D_1 \lambda_1 / U_2 (T_1 - T_2') = 0 \cdot 724 \times 2266 / [1 \cdot 1(370 - 339)] = 48 \cdot 0 \, \text{m}^2$$

which are close enough for design purposes.

The area to be specified for each effect would be $\underline{\underline{47 \cdot 5 \, \text{m}^2}}$.

Problem 14.25

2·5 kg/s of a solution at 288 K containing 10% of dissolved solids is fed to a forward-feed double-effect evaporator operating at a pressure of 14 kN/m² in the last effect. If the product is to consist of a liquid containing 50% by weight of dissolved solids and dry saturated steam is fed to the steam coils, what should be the pressure of the steam? The surface in each effect is 50 m² and the coefficients for heat transfer in the first and second effects are 2·8 and 1·7 kW/m² K respectively. Assume that the concentrated solution exhibits a boiling-point rise of 5 degK, that the latent heat has a constant value of 2260 kJ/kg, and that the specific heat of the liquid stream is constant at 3·75 kJ/kg K.

Solution

A mass balance gives:

	Solids (kg/s)	Liquid (kg/s)	Total (kg/s)
Feed	0·25	2·25	2·50
Product	0·25	0·25	0·50
Evaporation	—	2·0	2·0

$\therefore$
$$D_1 + D_2 = 2{\cdot}0\,\text{kg/s}$$

At $14{\cdot}0\,\text{kN/m}^2$,
$$T_2 = 326\,\text{K}$$

and allowing for the boiling-point rise in the second effect,
$$T_2' = 331\,\text{K}$$

Writing a heat balance for each effect:

(1) $D_0 \lambda_0 = W(T_1 - T_f) + D_1 \lambda_1$ or $2260 D_0 = 2{\cdot}5 \times 3{\cdot}75(T_1 - 288) + 2260 D_1$ (i)

(2) $D_1 \lambda_1 + (W - D_1)(T_1 - T_2') = D_2 \lambda_2$

    or $2260 D_1 + 2{\cdot}5 - D_1 + (2{\cdot}5 - D_1)3{\cdot}75(T_1 - 331) = 2260 D_2$

since $D_2 = 2{\cdot}0 - D_1$

$\therefore$
$$4520(1 - D_1) = (9{\cdot}375 - 3{\cdot}75 D_1)(T_1 - 331) \qquad\qquad \text{(ii)}$$

But
$$D_1 \lambda_1 = U_2 A_2 (T_1 - T_2')$$

$\therefore$
$$2260 D_1 = 1{\cdot}7 \times 50(T_1 - 331) \quad \text{or} \quad T_1 = 26{\cdot}6 D_1 + 331$$

Substituting in (ii) for T_1:
$$D_1^2 - 48{\cdot}26 D_1 + 45{\cdot}31 = 0$$

and $D_1 = 47{\cdot}30\,\text{kg/s}$ (which is clearly impossible) or $0{\cdot}96\,\text{kg/s}$

$\therefore$
$$T_1 = (26{\cdot}6 \times 0{\cdot}96) + 331$$
$$= 356{\cdot}5\,\text{K}$$

$\therefore$ In (i) $2260 D_0 = 2{\cdot}5 \times 3{\cdot}75(356{\cdot}5 - 288) + 2260 \times 0{\cdot}96$

and $D_0 = 1{\cdot}24\,\text{kg/s}$

But $D_0 \lambda_0 = U_1 A_1 (T_0 - T_1)$

or $1{\cdot}24 \times 2260 = 2{\cdot}8 \times 50(T_0 - 356{\cdot}5)$

$\therefore$ $T_0 = 376{\cdot}5\,\text{K}$

Steam is dry and saturated at $376{\cdot}5\,\text{K}$ at a pressure of $\underline{115\,\text{kN/m}^2}$.

Chemical Engineering

Problem 14.26

A salt solution at 293 K is fed at the rate of 6·3 kg/s to a forward-feed triple-effect evaporator and is concentrated from 2% to 10% of solids. Saturated steam at 170 kN/m² is introduced into the calandria of the first effect and a pressure of 34 kN/m² is maintained on the last effect. If the heat transfer coefficients in the three effects are respectively 1·7, 1·4, and 1·1 kW/m² K and the specific heat of the liquid is approximately 4 kJ/kg K, what area is required if each effect is identical? Condensate may be assumed to leave at the vapour temperature at each stage and the effects of boiling-point may be neglected. The latent heat of vaporisation may be taken as constant throughout.

Solution

The mass balance is as follows:

	Solids (kg/s)	Liquor (kg/s)	Total (kg/s)
Feed	0·126	6·174	6·30
Product	0·126	1·134	1·26
Evaporation	—	5·04	5·04

$$(D_1 + D_2 + D_3) = 5.04 \text{ kg/s}$$

At 170 kN/m², $T_0 = 388 \text{ K}$ and $\lambda_0 = 2216 \text{ kJ/kg}$

At 34 kN/m², $T_3 = 345 \text{ K}$ and $\lambda_3 = 2328 \text{ kJ/kg}$

Thus λ will be taken as 2270 kJ/kg throughout and

$$\Sigma\Delta T = (388 - 345) = 43 \text{ degK}$$

From equation 14.8:

$$1.7\Delta T_1 = 1.4\Delta T_2 = 1.1\Delta T_3$$

and hence $\Delta T_1 = 11.5 \text{ degK}$, $\Delta T_2 = 14 \text{ degK}$,   and $\Delta T_3 = 17.5 \text{ degK}$

Modifying these values for the cold feed, it will be assumed that:

$$\Delta T_1 = 15 \text{ degK}, \quad \Delta T_2 = 12 \text{ degK}, \quad \text{and} \quad \Delta T_3 = 16 \text{ degK}$$

Thus $T_0 = 388 \text{ K}$, $T_1 = 373 \text{ K}$, $T_2 = 361 \text{ K}$,   and $T_3 = 345 \text{ K}$

$$\lambda_0 = \lambda_1 = \lambda_2 = \lambda_3 = 2270 \text{ kJ/kg}$$

The heat balance over each effect is now:

(1) $D_0 \lambda_0 = W C_p(T_1 - T_f) + D_1 \lambda_1$ or $2270 D_0 = 6.3 \times 4(373 - 293) + 2270 D_1$

(2) $D_1 \lambda_1 + (W - D_1) C_p(T_1 - T_2) = D_2 \lambda_2$

or $2270 D_1 + (6.3 - D_1) 4(373 - 361) = 2270 D_2$

(3) $D_2 \lambda_2 + (W - D_1 - D_2) C_p(T_1 - T_2) = D_3 \lambda_3$

or $2270 D_2 + (6.3 - D_1 - D_2) 4(361 - 345) = 2270 D_3$

Solving:

$$D_0 = 2.48 \text{ kg/s}, \quad D_1 = 1.59 \text{ kg/s}, \quad D_2 = 1.69 \text{ kg/s}, \quad D_3 = 1.78 \text{ kg/s}$$

The areas are given by:

$$A_1 = D_0 \lambda_0 / U_1 \Delta T_1 = 2270 \times 2.48/(1.7 \times 15) = 220.8 \text{ m}^2$$
$$A_2 = D_1 \lambda_1 / U_2 \Delta T_2 = 2270 \times 1.59/(1.4 \times 12) = 214.8 \text{ m}^2$$
$$A_3 = D_2 \lambda_2 / U_3 \Delta T_3 = 2270 \times 1.69/(1.1 \times 16) = 217.9 \text{ m}^2$$

which are close enough for design purposes.

The area to be specified for each stage is therefore $\underline{218 \text{ m}^2}$.

Problem 14.27

A single-effect evaporator with 10 m^2 of heating surface is used to concentrate NaOH solution from 10% to 33.33%, the mass of entering feed being 0.38 kg/s. The feed enters at 338 K; its specific heat is 3.2 kJ/kg K. The pressure in the vapour spaces is 13.5 kN/m^2. 0.3 kg/s of steam is used from a supply at 375 K. Calculate:

(a) the apparent overall heat transfer coefficient;
(b) the coefficient corrected for boiling rise of dissolved solids;
(c) the corrected coefficient if the depth of liquid is 1.5 m.

Solution

Mass of solids in feed $= 0.38 \times 10/100 = 0.038 \text{ kg/s}$.
Mass flow of product $= 0.038 \times 100/33.33 = 0.114 \text{ kg/s}$.

$\therefore$ evaporation, $\qquad D_1 = (0.38 - 0.114) = 0.266 \text{ kg/s}$

At a pressure of 13.5 kN/m^2, steam is saturated at 325 K and $\lambda_1 = 2376 \text{ kJ/kg}$.
At 375 K, steam is saturated at 413 K and $\lambda_0 = 2140 \text{ kJ/kg}$.

(a) Ignoring any boiling-point rise, it may be assumed that the temperature of the boiling liquor, $T_1 = 325 \text{ K}$.

$$\therefore \qquad \Delta T_1 = 375 - 325 = 50 \text{ degK}$$
$$U_1 = D_0 \lambda_0 / A_1 \Delta T_1$$
$$= 0.3 \times 2140/(10 \times 50)$$
$$= \underline{1.28 \text{ kW/m}^2 \text{ K}}$$

(b) Allowing for a boiling-point rise, the temperature of the boiling liquor in the effect T_1' may be calculated from a heat balance:

$$D_0 \lambda_0 = W C_p (T_1' - T_f) + D_1 \lambda_1$$
$$\therefore \qquad 0.3 \times 2140 = 0.38 \times 3.2(T_1' - 338) + 0.266 \times 2376$$

and $T_1' = 346 \, \text{K}$

$$\Delta T_1 = 375 - 346 = 29 \, \text{degK}$$

and $U_1 = 0.3 \times 2140/(10 \times 29)$

$$= 2.21 \, \text{kW/m}^2 \, \text{K}$$

(c) Taking the density of the fluid as $1000 \, \text{kg/m}^3$, the pressure due to a height of liquid of $1.5 \, \text{m} = 1.5 \times 1000 \times 9.81$

$$= 14{,}715 \, \text{N/m}^2 \quad \text{or} \quad 14.7 \, \text{kN/m}^2$$

The pressure at the tubes is therefore $13.5 + 14.7 = 28.2 \, \text{kN/m}^2$ at which pressure, water boils at 341 K.

∴ $\Delta T_1 = 375 - 341 = 34 \, \text{degK}$

and $U_1 = 0.3 \times 2140/(10 \times 34)$

$$= 1.89 \, \text{kW/m}^2 \, \text{K}$$

[A value of the boiling liquor temperature $T_1' = 346 \, \text{K}$ obtained in (b) by heat balance must represent the effect of hydrostatic head *and* boiling-point rise. The true boiling-point rise must therefore be $(346 - 341) = 5 \, \text{degK}$.

∴ $T_1' = 325 + 5 = 330 \, \text{K}$

$$T_1 = 375 - 330 = 45 \, \text{K}$$

and $U_1 = 0.3 \times 2140/(10 \times 45) = 1.43 \, \text{kN/m}^2 \, \text{K}.]$

Problem 14.28

An evaporator, working at atmospheric pressure, is to concentrate a solution from 5% to 20% solids at the rate of $1.25 \, \text{kg/s}$. The solution which has a specific heat of $4.18 \, \text{kJ/kg K}$ is fed to the evaporator at 295 K and boils at 380 K. Dry saturated steam at $240 \, \text{kN/m}^2$ is fed to the calandria and the condensate leaves at the temperature of the condensing steam. If the heat transfer coefficient is $2.3 \, \text{kW/m}^2 \, \text{K}$, what is the required area of heat transfer surface and how much steam is required? The latent heat of vaporisation may be taken as the same as that of water.

Solution

A material balance in kg/s becomes:

	Solids (kg/s)	Liquor (kg/s)	Total (kg/s)
Feed	0.0625	1.1875	1.2500
Product	0.0625	0.2500	0.3125
Evaporation	—	0.9375	$0.9375 = D_1$

At 240 kN/m², $\qquad T_0 = 399\,\text{K}$ and $\lambda_0 = 2185\,\text{kJ/kg}$

At atmospheric pressure, 101·3 kN/m²,

$$\lambda_1 = 2257\,\text{kJ/kg}$$

Making a heat balance across the unit:

$$D_1\lambda_1 + WC_p(T_1 - T_f) = D_0\lambda_0$$

or $\qquad (2257 \times 0{\cdot}9375) + (1{\cdot}25 \times 4{\cdot}18)(380 - 295) = 2185D_0$

$\therefore \qquad\qquad\qquad \underline{D_0 = 1{\cdot}17\,\text{kg/s}}$

The heat transfer area is given by:

$$A = D_0\lambda_0/UT_1$$
$$= 1{\cdot}17 \times 2185/[2{\cdot}3(399 - 380)]$$
$$= \underline{\underline{58{\cdot}5\,\text{m}^2}}$$

SECTION 15

CRYSTALLISATION

Problem 15.1

A saturated solution containing 1500 kg of potassium chloride at 360 K is cooled in an open tank to 290 K. If the specific gravity of the solution is 1·2, the solubility of potassium chloride per 100 parts of water is 53·55 at 360 K and 34·5 at 290 K; calculate:

(a) the capacity of the tank required;
(b) the weight of crystals obtained, neglecting loss of water by evaporation.

Solution

Assuming the solubility data are in terms of mass, at 360 K, 1500 kg KCl will be dissolved in $(1500 \times 100)/53{\cdot}55 = 2801$ kg water.

The total mass of the solution $= (1500 + 2801) = 4301$ kg.

The density of the solution $= (1{\cdot}2 \times 1000) = 1200 \, \text{kg/m}^3$ and hence the capacity of the tank $= (4301/1200) = 3{\cdot}58 \, \text{m}^3$.

At 290 K, mass of KCl dissolved in 2801 kg water

$$= (2801 \times 34{\cdot}5/100) = 966 \, \text{kg}$$

∴ mass of crystals which has come out of solution

$$= (1500 - 966)$$

$$= \underline{534 \, \text{kg}}$$

Problem 15.2

What do you understand by the term 'fractional crystallisation'? Explain with the aid of a diagram how the operation can be applied to a mixture of sodium chloride and sodium nitrate given the following data. At 293 K the solubility of sodium chloride is 36 parts per 100 of water and of sodium nitrate 88 parts per 100 of water. Whilst at this temperature, a saturated solution comprising both salts will contain 25 parts of sodium chloride and 59 parts of sodium nitrate per 100 parts of water. At 373 K these figures per 100 parts of water are 40, 176 and 17, 160, respectively.

Solution

The operation of fractional crystallisation is discussed in Vol. 2, Section 15.2.6.

The data given enable a plot of parts NaCl per 100 parts water to be drawn against parts NaNO₃ per 100 parts water as shown in Fig. 15a. On the diagram, points A and B represent solutions saturated with respect to both NaCl and NaNO₃ at 293 K and 373 K respectively. Fractional crystallisation may be applied to this system as follows:

(i) A solution saturated with both NaCl and NaNO₃ is made up at 373 K. This is represented by point B, and on the basis of 100 parts water this contains 17 parts NaCl and 160 parts NaNO₃.

(ii) The solution is separated from any residual solid and then cooled to 293 K. In so doing, the composition of the solution moves along BD.

(iii) At D this point lies on AC, which represents solutions saturated with NaNO₃ but not with NaCl. Thus the solution still contains 17 parts NaCl and in addition is saturated with 68 parts of NaNO₃. That is $(160 - 68) = 92$ parts of pure NaNO₃ crystals have come out of solution and may be drained and washed.

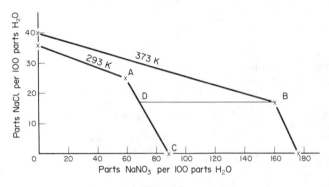

FIG. 15a

In this way, relatively pure NaNO₃, depending on the choice of conditions and particle size, has been separated from a mixture of NaNO₃ and NaCl.

The per cent NaNO₃ recovered from the saturated solution at 373 K

$$= (92 \times 100)/160 = \underline{\underline{57 \cdot 5\%}}$$

SECTION 16

DRYING

Problem 16.1

A wet solid is dried from 35% to 10% moisture under constant drying conditions in 18 ks (5 h). If the equilibrium moisture content is 4% and the critical moisture content is 14%, how long will it take to dry 6% moisture under the same conditions?

Solution

For the first set of conditions,

$$w_1 = 0.35 \text{ kg/kg}, \qquad w_e = 0.04 \text{ kg/kg}$$
$$w = 0.10 \text{ kg/kg}, \qquad w_c = 0.14 \text{ kg/kg}$$

$\therefore$
$$f_1 = (w_1 - w_e) = 0.31 \text{ kg/kg}$$
$$f_c = (w_c - w_e) = 0.10 \text{ kg/kg}$$
$$f = (w - w_e) = 0.06 \text{ kg/kg}$$

and from equation 16.14 the total drying time is:

$$t = (1/mA)[\ln(f_c/f) + (f_1 - f_c)/f_c]$$

or
$$18 = (1/mA)[\ln(0.10/0.06) + (0.31 - 0.10)/0.10]$$

$\therefore$
$$mA = 0.055(0.511 + 2.1) = 0.145 \text{ kg/s}$$

For the second set of conditions:

$$w_1 = 0.35 \text{ kg/kg}, \qquad w_c = 0.04 \text{ kg/kg}$$
$$w = 0.06 \text{ kg/kg}, \qquad w_c = 0.14 \text{ kg/kg}$$

$\therefore$
$$f_1 = (0.35 - 0.04) = 0.31 \text{ kg/kg}$$
$$f_c = (0.14 - 0.04) = 0.10 \text{ kg/kg}$$
$$f = (0.06 - 0.04) = 0.02 \text{ kg/kg}$$

and in equation 16.14, taking mA as 0.145 kg/s, the total drying time,

$$t = (1/0.145)[\ln(0.10/0.02) + (0.31 - 0.10)/0.10]$$
$$= 6.89(1.609 + 2.1)$$
$$= 25.6 \text{ ks } (7.1 \text{ h})$$

Problem 16.2

Strips of material 10 mm thick are dried under constant drying conditions from 28% to 13% moisture in 25 ks (7 h). If the equilibrium moisture content is 7%, find the time taken to dry 60 mm planks from 22% to 10% moisture under the same conditions assuming no loss from the edges. All moistures are given on the wet basis.

The relation between E, the ratio of the average free moisture content at time t to the initial free moisture content, and the parameter j is given by:

E	1	0·64	0·49	0·38	0·295	0·22	0·14
j	0	0·1	0·2	0·3	0·5	0·6	0·7

Note that $j = kt/l^2$, where k is a constant, t the time in ks and $2l$ the thickness of the sheet of material in millimetres.

Solution

For the 10 mm strips

Initial free moisture content $= (0·28 - 0·07) = 0·21\,\text{kg/kg}$.
Final free moisture content $= (0·13 - 0·07) = 0·06\,\text{kg/kg}$.

$\therefore$ when $t = 25\,\text{ks}$, $\qquad\qquad E = 0·06/0·21 = 0·286$

and from Fig. 16a a plot of the given data,

$$j = 0·52$$

$\therefore$
$$0·52 = k \times 25/(10/2)^2$$

$\therefore$
$$k = 0·52$$

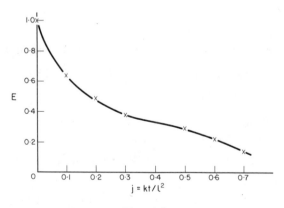

FIG. 16a

For the 60 mm planks

Initial free moisture content $= (0·22 - 0·07) = 0·15\,\text{kg/kg}$.
Final free moisture content $= (0·10 - 0·07) = 0·03\,\text{kg/kg}$.

$$E = 0.03/0.15 = 0.20$$

From Fig. 16a, $j = 0.63$

and hence $t = jl^2/k$

$$= 0.63 \times (60/2)^2/0.52$$

$$= \underline{1090\,ks} \quad (12.6\,days)$$

Problem 16.3

A granular material containing 40% moisture is fed to a countercurrent rotary dryer at a temperature of 295 K and is withdrawn at 305 K containing 5% moisture. The air supplied, which contains 0·006 kg water vapour per kg of dry air, enters at 385 K and leaves at 310 K. The dryer handles 0·125 kg/s wet stock.

Assuming that radiation losses amount to 20 kJ/kg dry air used, determine the weight of dry air supplied to the dryer per second and the humidity of the air leaving it.

Latent heat of water vapour at 295 K = 2449 kJ/kg.
Specific heat of dried material = 0·88 kJ/kg K.
Specific heat of dry air = 1·00 kJ/kg K.
Specific heat of water vapour = 2·01 kJ/kg K.

Solution

This problem is concerned with a heat balance over the system. 273 K will be chosen as the datum temperature and it will be assumed that the flow of dry air = m kg/s.

Heat in

(i) *Air*

m kg/s dry air enter with 0·006m kg/s water vapour and hence heat in this stream:

$$= [(1.00m) + (0.006m \times 2.01)](385 - 273)$$

$$= 113.35m\,kW$$

(ii) *Wet solid*

0·125 kg/s enter containing 0·40 kg water/kg wet solid (assuming moisture on a wet basis).

∴   mass flow of water $= (0.125 \times 0.40) = 0.050$ kg/s

and   mass flow of dry solid $= (0.125 - 0.050) = 0.075$ kg/s

Hence heat in this stream

$$= [(0.050 \times 4.18) + (0.075 \times 0.88)](295 - 273)$$

$$= 6.05\,kW$$

Heat out

(i) *Air*

Heat in exit air $= [(1{\cdot}00m) + (0{\cdot}006m \times 2{\cdot}01)](310 - 273) = 37{\cdot}45m\,$kW.
Mass flow of dry solids $= 0{\cdot}075\,$kg/s containing $0{\cdot}05\,$kg water/kg wet solids.
Hence water in dried solids leaving $= (0{\cdot}05 \times 0{\cdot}075)/(1 + 0{\cdot}05)$

$$= 0{\cdot}0036\,\text{kg/s}$$

and water evaporated into gas steam $= (0{\cdot}050 - 0{\cdot}0036) = 0{\cdot}0464\,$kg/s.

Assuming evaporation takes place at 295 K,

$$\text{heat in water vapour} = 0{\cdot}0464[2{\cdot}01(310 - 295) + 2449 + 4{\cdot}18(295 - 273)]$$

$$= 119{\cdot}30\,\text{kW}$$

and total heat in this stream $= (119{\cdot}30 + 37{\cdot}45m)\,$kW.

(ii) *Dried solids*

The dried solids contain $0{\cdot}0036\,$kg/s water and hence heat content of this stream

$$= [(0{\cdot}075 \times 0{\cdot}88) + (0{\cdot}0036 \times 4{\cdot}18)/(305 - 273)]$$

$$= 2{\cdot}59\,\text{kW}$$

(iii) *Losses*

These amount to $20\,$kJ/kg dry air or $20m\,$kW.

Balance

$$113{\cdot}35m + 6{\cdot}05 = 119{\cdot}30 + 37{\cdot}45m + 2{\cdot}59 + 20m$$

$$\therefore \qquad m = \underline{2{\cdot}07\,\text{kg/s}}$$

Water in outlet air stream $= (0{\cdot}006 \times 2{\cdot}07) + 0{\cdot}0464 = 0{\cdot}0588\,$kg/s

and humidity $= 0{\cdot}0588/2{\cdot}07 = \underline{0{\cdot}0284\,\text{kg/kg dry air}}$

Problem 16.4

One megagram (1 tonne) (dry weight) of a non-porous solid is dried under constant drying conditions with an air velocity of $0{\cdot}75$ m/s, so that the area of surface drying is 55 m^2. If the initial rate of drying is $0{\cdot}3$ g/m^2 s, how long will it take to dry the material from $0{\cdot}15$ to $0{\cdot}025$ kg water/kg dry solid? The critical moisture content of the material may be taken as $0{\cdot}125$ kg water/kg dry solid.

If the air velocity was increased to $4{\cdot}0$ m/s, what would be the anticipated saving in time if surface evaporation is controlling?

Solution

During the constant rate period, that is whilst the moisture content falls from 0·15 to 0·125 kg/kg, the rate of drying,

$$\frac{1}{A}\frac{dw}{dt} = 0·3 \times 10^{-3}\,\text{kg/m}^2\,\text{s}$$

At the start of the falling rate period,

$$w = w_c = 0·125\,\text{kg/kg}$$

and

$$\frac{1}{A}\frac{dw}{dt} = m(w_c - w_e)$$

$$0·3 \times 10^{-3} = m(0·125 - 0·025)$$

or

$$m = 3·0 \times 10^{-3}\,\text{kg/m}^2\,\text{s kg dry solid}$$

$$= 3·0 \times 10^{-6}\,\text{kg/m}^2\,\text{s Mg dry solid}$$

The total drying time is given by equation 16.14:

$$t = (1/mA)[\ln(f_c/f) + (f_1 - f_c)/f_c]$$

where

$$f = (0·025 - 0) = 0·025\,\text{kg/kg} \quad (\text{taking } w_e = 0)$$

$$f_c = (0·125 - 0) = 0·125\,\text{kg/kg}$$

$$f_1 = (0·15 - 0) = 0·150\,\text{kg/kg}$$

∴

$$t = [1/(3·0 \times 10^{-6} \times 55)][\ln(0·125/0·025) + (0·150 - 0·125)/0·125]$$

$$= 10·96 \times 10^3\,\text{s}$$

or

$$\underline{\underline{10·96\,\text{ks} \quad (3·04\,\text{h})}}$$

As a first approximation it may be assumed that the rate of evaporation is proportional to the air velocity raised to the power of 0·8. Hence for the second case, *m* may be calculated as:

$$m = 3·0 \times 10^{-6}(4·0/0·75)^{0·8}$$

$$= 1·15 \times 10^{-5}\,\text{kg water/m}^2\,\text{s Mg dry solid}$$

Hence the time of drying is:

$$t = [1/(1·15 \times 10^{-5} \times 55)](1·609 + 0·20)$$

$$= 2·86 \times 10^3\,\text{s} \quad \text{or} \quad 2·86\,\text{ks}$$

The time saved is therefore $\underline{\underline{8·10\,\text{ks} \quad (2·25\,\text{h})}}$.

Problem 16.5

A 100 kg batch of granular solids containing 30% moisture is to be dried in a tray drier to 15·5% of moisture by passing a current of air at 350 K tangentially across its surface at a velocity of 1·8 m/s. If the constant rate of drying under these conditions is 0·7 g/s m² and the critical moisture content is 15%, calculate approximately the drying time.

Assume the drying surface to be 0·03 m²/kg dry weight.

Solution

In 100 kg feed, weight of water = (100 × 30/100) = 30 kg

and weight of dry solids = (100 − 30) = 70 kg

For x kg water in dried solids:

$$100x/(x + 70) = 15.5$$

$$100x = 15.5 + 1085$$

and water in product $x = 12.8$ kg

∴ initial moisture content, w_1 = 30/70 = 0·429 kg/kg dry solids

final moisture content, w_2 = 12·8/70 = 0·183 kg/kg dry solids

and water to be removed = (30 − 12·8) = 17·2 kg

The surface area available for drying = (0·03 × 70) = 2·1 m² and hence the rate of drying during the constant period = 0·7 × 2·1 = 1·47 g/s or 0·001 47 kg/s.

As the final moisture content is above the critical value, all the drying is at this constant rate and the time of drying,

$$t = (17.2/0.001\ 47) = 11,700\ \text{s}$$

or 11·7 ks (3·25 h)